Spectroscopic Membrane Probes

Volume II

Editor

Leslie M. Loew, Ph.D.
Professor of Physiology
University of Connecticut Health Center
Farmington, Connecticut

CRC Press
Taylor & Francis Group
Boca Raton London New York

CRC Press is an imprint of the
Taylor & Francis Group, an **informa** business

First published 1988 by CRC Press
Taylor & Francis Group
6000 Broken Sound Parkway NW, Suite 300
Boca Raton, FL 33487-2742

Reissued 2018 by CRC Press

© 1988 by Taylor & Francis
CRC Press is an imprint of Taylor & Francis Group, an Informa business

No claim to original U.S. Government works

Publisher's Note
The publisher has gone to great lengths to ensure the quality of this reprint but points out that some imperfections in the original copies may be apparent.

Disclaimer
The publisher has made every effort to trace copyright holders and welcomes correspondence from those they have been unable to contact.

ISBN 13: 978-1-315-89769-1 (hbk)
ISBN 13: 978-1-351-07679-1 (ebk)

Visit the Taylor & Francis Web site at http://www.taylorandfrancis.com and the
CRC Press Web site at http://www.crcpress.com

DEDICATION

To my parents.

ACKNOWLEDGMENTS

I am pleased to acknowledge my wife, Helen, for putting up with late-night word processing in our bedroom and for her selfless support throughout my career.

I thank those contributors to this work who were able to provide their manuscripts within 6 months after the deadline. My thanks are offered not only for their punctuality but also for their tolerance of their more tardy colleagues. I also wish to thank all of the contributors for their willingness to work with me on this project and for the high quality of each of their chapters.

PREFACE

The optical spectra of molecular membrane probes can be interpreted in terms of the structure, dynamics, and physiological state of the membrane. The general picture we have of membranes and of the properties of the proteins imbedded in them, has, arguably, emerged directly from probe studies over the last 20 years. This work is designed to make these techniques accessible to a broad audience of cell biologists.

The techniques discussed revolve primarily around the fluorescence of membrane probes, but applications of light absorption and Raman scattering are included. In addition to reviews of the major applications, most chapters include information on the required apparatus, experimental design, data analysis, and potential pitfalls. Thus, significant attention has been given to practical considerations involved in the use of these methods. The aim is to provide the novice with some appreciation of the real requirements, with respect to both expertise and equipment, for the implementation of these methods.

Specific methods that are represented include fluorescence anisotropy, resonance energy transfer, fluorescence recovery after photobleaching, digital video microscopy, total internal reflection fluorescence, near-field imaging, multisite optical recordings of electrical activity, and resonance Raman spectroscopy. Chapter 1 gives an overview of molecular photophysics. It serves to introduce Chapters 2 to 13 which cover dynamic membrane structure, membrane fusion assays, and studies of the properties of specific membrane proteins. Chapter 14 is a practical and concise guide to the characteristics of potentiometric membrane dyes and introduces Chapters 15 to 22 which describe specific applications. These include measurements of membrane potential in preparations ranging from isolated energy-transducing organelles, through cell suspensions and individual cells, to complex neuronal systems. The result is a most comprehensive examination of probe methods for the study of membrane potential, which can serve as a convenient source for the design of new physiological applications of these dyes.

THE EDITOR

Leslie M. Loew, Ph.D., is Professor of Physiology at the University of Connecticut Health Center in Farmington.

Dr. Loew received his B.S. degree from the City College of the City University of New York in 1969. Under the tutelage of Professor C. F. Wilcox, he obtained his M.S. and Ph.D. degrees in 1972 and 1974, respectively, from the Department of Chemistry, Cornell University, Ithaca, N.Y. After a year's postdoctoral training with F. H. Westheimer at Harvard University, he joined the Department of Chemistry at the State University of New York at Binghamton in 1974 as Assistant Professor. He became an Associate Professor of Chemistry in 1979 at SUNY Binghamton. As a result of a steady shift in research interest from physical organic chemistry to membrane biophysics, Dr. Loew moved to the University of Connecticut, where he was appointed Associate Professor of Physiology in 1984 and Professor in 1987. He has retained his association with SUNY Binghamton by holding the position of Adjunct Associate Professor of Chemistry.

Dr. Loew is a member of the American Association for the Advancement of Science, the American Chemical Society, the Biophysical Society, and Phi Beta Kappa. He was a Research Career Development Awardee of the National Institutes of Health (1980 to 1985) and has been a visiting faculty member at Cornell University (1984), Bar Ilan University (1983), and the Weizmann Institute of Science (1981, 1983, 1985).

He has been the recipient of research grants from the National Institutes of Health, the American Chemical Society, the Research Corporation, and private industry. He has published more than 50 papers. His current research interests are in the design of potentiometric membrane dyes and the study of the electrical properties of membranes.

CONTRIBUTORS

Volume II

Daniel Axelrod, Ph.D.
Professor
Department of Physics and Biophysics
 Research Division
University of Michigan
Ann Arbor, Michigan

Eric Betzig
School of Applied and Engineering
 Physics
Cornell University
Ithaca, New York

James A. Dix, Ph.D.
Associate Professor
Department of Chemistry
State University of New York
Binghamton, New York

Elliot L. Elson, Ph.D.
Department of Biological Chemistry
Washington University School of
 Medicine
St. Louis, Missouri

Robert M. Fulbright
Research Associate
Department of Physics and Biophysics
 Research Division
University of Michigan
Ann Arbor, Michigan

David J. Gross, Ph.D.
Research Associate
School of Applied and Engineering
 Physics
Cornell University
Ithaca, New York

Alec Harootunian, Ph.D.
School of Applied and Engineering
 Physics
Cornell University
Ithaca, New York

Edward H. Hellen
Research Associate
Department of Physics and Biophysics
 Research Division
University of Michigan
Ann Arbor, Michigan

Michael Isaacson, Ph.D.
Associate Professor
School of Applied and Engineering
 Physics
Cornell University
Ithaca, New York

Ernst Kratschmer, Ph.D.
National Research and Resource Facility
 for Submicron Structures
Cornell University
Ithaca, New York

Aaron Lewis, Ph.D.
Professor
School of Applied and Engineering
 Physics
Cornell University
Ithaca, New York

Jerry C. Smith, Ph.D.
Associate Professor
Department of Chemistry and LMBS
Georgia State University
Atlanta, Georgia

Watt W. Webb, Sc.D.
Professor and Director
School of Applied and Engineering
 Physics
Cornell University
Ithaca, New York

TABLE OF CONTENTS

Volume I

Volume II

Volume III

Chapter 9

FLUORESCENCE CORRELATION SPECTROSCOPY AND PHOTOBLEACHING RECOVERY: MEASUREMENT OF TRANSPORT AND CHEMICAL KINETICS

Elliot L. Elson

TABLE OF CONTENTS

I. INTRODUCTION

Fluorescence Correlation Spectroscopy (FCS) and Fluorescence Photobleaching Recovery (FPR) are two closely related methods for measuring chemical kinetics and rates of molecular transport, including diffusion and systematic flow or drift. They are similar in concept and experimental implementation. Both methods are based on measurements over time of the fluorescence intensity emitted from a defined open subregion of the sample. The intensity is proportional to the numbers of the fluorescent molecules in the subregion. These numbers change, due either to chemical reactions or to transport into or out of the region. Hence, measurements of rates of change of fluorescence, and thereby rates of change of numbers of molecules specify the desired transport and chemical reaction rates.

FCS and FPR differ, however, in certain crucial respects. FCS measures fluorescence changes that result from spontaneous fluctuation of the number of fluorophores in the sampled subregion while the system rests in thermodynamic equilibrium. To determine accurately the phenomenological coefficients for transport and chemical kinetics it is necessary to measure and analyze statistically many fluorescence fluctuations. [The "phenomenological" diffusion coefficients and chemical kinetic constants are those normally used to describe rates of transport and reaction in systems containing many molecules which undergo macroscopic changes over relatively long periods of time.[1]] This statistical treatment is needed not only because the fluctuations are typically small and difficult to measure precisely. In addition, the time course of an individual fluctuation is related only probabilistically to the phenomenological coefficients.[1] Hence, even if individual fluctuations could be measured with infinite precision, a statistical analysis of many fluctuations would be necessary. Therefore, conventional FCS measurements often require long periods for accumulation of data. In contrast, FPR measures the much larger changes in fluorescence which arise from macroscopic concentration gradients generated by photobleaching a fraction of the fluorescence in the observed subregion. Because the concentration gradient is macroscopic, its time course is accurately governed by the phenomenological coefficients, which can therefore be determined from measurements of single fluorescence recovery transients.

Hence, the principal experimental difference between FCS and FPR is in the size and number of the fluorescence transients that must be recorded. Therefore, the apparatus, typically based on a fluorescence microscope with a laser light source, is very similar for the two kinds of measurements.[2,3] Beyond the common features, however, FPR requires specialized electronics and optics for controlling the photobleaching pulse, and FCS measurements demand the capability for statistical analysis of the large number of observed fluorescence fluctuations.

A specific advantage in principle of FCS over FPR is that, since the system rests in thermodynamic equilibrium throughout the measurement, it is not necessary to employ and characterize a process which generates an initial nonequilibrium state. Uncertainties about the mechanism of photobleaching can lead to uncertainties in the detailed interpretation of FPR experiments. This is especially true for systems in which many fluorophores behave as a single kinetic unit as in multiple binding or polymerization equilibria.[4-6] A substantial practical disadvantage of FCS results from the need to accumulate many tiny fluctuation transients, each of which requires about the same time as a single macroscopic fluorescence recovery in an FPR measurement. Hence, ordinarily FCS can be applied only to very stable systems. The stability of living cells might be compromised under prolonged exposure to the measurement conditions, including the excitation light. Even in the absence of physiological perturbations, however, systematic cellular motions such as ruffling, extension of microvilli, and locomotion could generate fluorescence changes which could overwhelm the small fluorescence signals due to spontaneous concentration fluctuations. In contrast, FPR enjoys the advantage that a single recovery transient is sufficient in principle to determine the phenomenological coefficients, although signal averaging is sometimes useful to improve accuracy. Hence, FPR has been extensively used in studies of cell surface mobility, while FCS has found few applications to living cells. Nevertheless, FCS can provide unique information about aggregation states of fluorophores not accessible to FPR. Furthermore, as discussed in the following pages, there are ways of performing FCS measurements which reduce the time needed for data accumulation (at some expense of information obtained) and which might therefore allow the extension of FCS to otherwise inaccessible problems.

This chapter discusses the implementation and interpretation of FCS and FPR measurements with some emphasis on the applicability of these methods to study chemical kinetics. For historical background and a discussion of applications see Reference 6 and the papers and reviews cited therein.

II. RELATION OF MEASUREMENTS TO PHENOMENOLOGICAL COEFFICIENTS

In an FPR experiment the photobleaching reaction generates a macroscopic gradient, $\Delta c(\mathbf{r},t) = c(\mathbf{r},t) - \bar{c}$, between the concentration, $c(\mathbf{r},t)$, at position $\mathbf{r}$ and time t in the interior of the observed region and $\bar{c}$, the equilibrium concentration, prior to the photobleaching pulse and in the unperturbed surrounding parts of the system (supposed to be the same in both). This is gauged by a corresponding macroscopic displacement of the measured fluorescence, $\Delta f(\mathbf{r},t) = f(\mathbf{r},t) - \bar{f}$. Here, $f(\mathbf{r},t)$, is the fluorescence per unit area of the sample at $\mathbf{r}$ and t and $\bar{f}$ is the prebleach and equilibrium value. Frequently, one measures not the local fluorescence displacement per unit area, but, rather the displacement of the total fluorescence from a defined region, $\Delta F(t)$, where $\Delta F(t) = \int \Delta f(\mathbf{r},t)d^2r$, and the integration extends over the region of interest. For simplicity we are assuming that the system is confined to a plane which coincides with the plane of focus of the detection optical system (typically a microscope). The measurement consists of observing the relaxation of $\Delta f(\mathbf{r},t)$ or $\Delta F(t)$ which results from the dissipation of the concentration gradient.

In an FCS experiment, however, with the system in equilibrium throughout the measurement, one observes the microscopic changes of fluorescence emitted from a sample region, $\delta F(t)$, that result from the spontaneous concentration fluctuations, $\delta c(\mathbf{r},t)$, and that occur in an open system even when in equilibrium. The required statistical analysis is usually accomplished by computing a fluorescence fluctuation autocorrelation function, $G(\tau)$,

$$G(\tau) = <\delta F(0)\delta F(\tau)> = \underset{\Xi \to \infty}{\text{Lim}} \; 1/\Xi \int_0^\Xi \delta F(t)\delta F(t + \tau)dt \tag{1}$$

where $<...>$ denotes an ensemble average and the second equation supplies an operational definition.[7,8] (We have supposed that the system is in equilibrium and therefore is stationary.)

The primary experimental datum is a measurement of the microscopic or macroscopic displacement of the fluorescence from its initial or equilibrium value. This displacement $\delta F(t)$ must be related to the displacements of the concentration of the fluorescent components of the system from their equilibrium values. The fluorescence emitted by component i from a position $\mathbf{r}$ at time t is proportional to the concentration of the component $c_i(\mathbf{r},t)$ weighted by the excitation intensity $I(\mathbf{r})$, and summed (integrated) over all positions:

$$\Delta f(\mathbf{r}, t) = \sum_i Q_i I(\mathbf{r})\Delta c_i(\mathbf{r}, t) \tag{2A}$$

$$\Delta F(t) = \sum_i Q_i \int I(\mathbf{r})\Delta c_i(\mathbf{r}, t)d^2r \tag{2B}$$

where Q_i accounts for the instrumental constants and for the absorption and fluorescence quantum yield of component i. Similarly,

$$G(\tau) = \sum_{i,j} Q_i Q_j \int I(\mathbf{r})I(\mathbf{r}') < \delta c_i(\mathbf{r}, 0)\delta c_j(\mathbf{r}', \tau) > d^2r d^2r' \qquad (3)$$

The dependence of the concentrations on position, time, and the phenomenological coefficients is determined from a system of differential equations which account for diffusion, systematic drift or flow, and chemical reaction:

$$\partial \delta c_i(\mathbf{r}, t)/\partial t = D_i \nabla^2 \delta c_i(\mathbf{r}, t) - V_i \partial \delta c_i(\mathbf{r}, t)/\partial x + \sum_j T_{ij}\delta c_j(\mathbf{r}, t) \qquad (4)$$

where D_i is the diffusion coefficient and V_i the uniform drift or flow velocity (arbitrarily taken to be in the x-direction) for component i, and the T_{ij} characterizes the contributions of component j to the kinetics of reactions involving component i. To use these equations to characterize the macroscopic concentration changes observed in FPR experiments we must assume that the displacements of concentrations from their equilibrium values are small enough so that the kinetic equations describing any nonlinear chemical reactions can be linearized. The systematic translation along the x-direction characterized by the constant velocity V_i could be due to drift in a field (e.g., electrophoresis or sedimentation), to uniform motion of the sample relative to the probe beam, or to a uniform flow process. Nonuniform flow fields, e.g., as in flow through a channel, can be much more difficult to analyze.[9]

These equations are most simply solved for systems in which the overall dimensions are large compared to the size of the observation area so that: $\delta c_i(\mathbf{r},t) = 0$ at $\mathbf{r} = (+/-)\infty$. The initial conditions are different for FCS and FPR experiments. For the former one supposes that the system behaves ideally, which is usually justifiable at the low concentrations at which these experiments are carried out. Then the number of molecules of each component in any volume will be independent of the concentration of any other component and will be governed by a Poisson distribution. Hence, the variance of the number of molecules (i.e., the mean square fluctuation) will equal the mean number: $<\delta c_i(\mathbf{r},0)\delta c_j(\mathbf{r}',0)> = \bar{c}_j \delta_{ij} \delta(\mathbf{r}-\mathbf{r}')$.[7] In an FPR experiment, the initial concentration gradient, $\Delta c_i(\mathbf{r},0)$, is set by the duration, T, and the intensity profile, $I'(\mathbf{r})$, of the photobleaching pulse and by the system of equations which describe the kinetics of the bleaching reaction. With these boundary and initial conditions one can conveniently solve the kinetic equations using a Fourier transform approach.[7] The solutions are expressed in terms of the eigenvalues, $\lambda^{(s)}$, and right and left eigenvectors, $X_j^{(s)}$ and $(X-1)_\ell^{(s)}$ of the matrix $M_{j\ell} = -[v^2 D_j \delta_{j\ell} - iv_x v_j] + T_{j\ell}$ where v is the Fourier transform variable.

As a simple example, one may consider a one-component system in which there is diffusion and drift along the x-direction (but no chemical reaction). Then,

$$\partial \delta c(\mathbf{r}, t)/\partial t = D\nabla^2 \delta c_i(\mathbf{r}, t) - V \partial \delta c(\mathbf{r}, t)/\partial x \qquad (5A)$$

and

$$c(\mathbf{v}, t) = c(\mathbf{v}, 0) \exp[-(v_x^2 + v_y^2)Dt + iV v_x t] \qquad (5B)$$

where

$$c(\mathbf{v}, 0) = (2\pi)^{-1} \int \exp(i\mathbf{v} \cdot \mathbf{r})\Delta c(\mathbf{r}, 0)d^2r$$

Hence the local fluorescence displacement, represented as an inverse Fourier transform, and the FCS correlation function are respectively,

$$\Delta f(\mathbf{r}, t) = QI(\mathbf{r})(2\pi)^{-1} \int \exp(-i\boldsymbol{\nu} \cdot \mathbf{r})c(\boldsymbol{\nu}, t)d^2\nu \qquad (5C)$$

and

$$G(\tau) = [Q^2c/(4\pi^2)] \int \exp[-(\nu^2D - iV\nu_x)\tau]\Phi(\boldsymbol{\nu})d^2\nu$$

where

$$\Phi(\boldsymbol{\nu}) = [(4\pi^2)^{-1} \int \exp(i\boldsymbol{\nu} \cdot \mathbf{r})I(\mathbf{r})d^2\mathbf{r}]^2$$

More generally:

$$G(\tau) = \sum_{j,\ell,s} Q_j Q_\ell c_\ell \int A_{j,\ell}^{(s)}\exp(\lambda^{(s)}\tau) \, \Phi(\boldsymbol{\nu})d^2\nu \qquad (6)$$

where: $A_{j,\ell} = X_j^{(s)}(X^{-1})_\ell^{(s)}$,[4,6] For an FPR experiment;

$$\Delta F(t) = \sum_{j,\ell,s} Q_j \int A_{j,\ell}^s\exp(\lambda^{(s)}\tau) \, \psi_\ell(\boldsymbol{\nu})d^2\nu \qquad (7)$$

where $\Psi_\ell(\boldsymbol{\nu}) = \int I(\mathbf{r})\exp(-i\boldsymbol{\nu}\cdot\mathbf{r})d^2\mathbf{r}\int\exp(i\boldsymbol{\nu}\cdot\mathbf{r}')\delta c_\ell(\mathbf{r}',0)d^2\mathbf{r}$.[4,6]

The evaluation of these expressions requires the specification of both the photochemical mechanism for the bleaching process and of the intensity profiles, $I(\mathbf{r})$ and $I'(\mathbf{r})$. In the simplest FPR experiments the photobleaching intensity profile is an intensified version of the monitoring intensity profile: $I'(\mathbf{r}) = BI(\mathbf{r})$ where the constant B is typically in the range 10^3 to 10^4. The simplest mechanistic assumption would hold that each fluorescent component of the system is irreversibly eliminated from the system independently of the others by a single first-order process.[10] Then,

$$dc_i(\mathbf{r}, t)/dt = -\gamma_i I'(\mathbf{r})c_i(\mathbf{r}, t) \qquad (8A)$$

so that:

$$\Delta c_i(\mathbf{r}, 0) = \bar{c}_i\{1 - \exp[-\gamma_i BI(\mathbf{r})T]\} \qquad (8B)$$

where γ_i is the photochemical rate constant. The assumption of an irreversible first order process may be only approximate and should be checked experimentally. Even if the individual bleaching events are irreversible and first order, the assumption that each of the components is eliminated independently of the others is inappropriate for some systems, especially those in which several fluorophores reside on a single kinetic unit, as in multiple binding and polymerization systems. Then, a more general model is required. If we continue to suppose that all photobleaching processes are first order in fluorophore concentrations, then we can represent the kinetics of a system of coupled reactions as:

$$dc_i(\mathbf{r}, t)/dt = I'(\mathbf{r}) \sum_j P_{ij}c_j(\mathbf{r}, t) \qquad (9)$$

Then a complete evaluation of Equation 7 would, in general, require the determination of the eigenvalues and eigenvectors of P_{ij} and the evaluation therefrom of the $\Delta c_i(\mathbf{r},0)$.

One practical difficulty in the application of FCS and FPR to multiple reaction systems is the complexity of the analysis of kinetic models, which requires the determination of the eigenvalues and eigenvectors of M_{ij}, and possibly P_{ij}, and then the integrations indicated in

Equations 6 and 7. In more conventional perturbation kinetics methods, e.g., the temperature-jump method, the analysis of the matrix T_{ij} presents a corresponding problem which becomes rapidly more complex as the number of reactions increases. The analysis of M_{ij} for FPR or FCS measurements on open systems can pose a problem of even higher dimension. Because the system is open, each component can change independently of the others (e.g., by independently diffusing into or out of the observation region). Hence, conservation conditions which can be used to reduce the number of independent components in a closed system do not hold for open systems. For example, a conventional relaxation kinetics measurement in a closed system of the bimolecular reaction, $A + B \rightarrow C$, can be analyzed in terms of a single relaxation mode with an amplitude set by the method of perturbation of the equilibrium state and a characteristic rate, $R = k_f(\bar{c}_A + \bar{c}_B) + k_b$, where k_f and k_b are the association and dissociation rate constants and $\bar{c}_A$ and $\bar{c}_B$ are the equilibrium concentrations of A and B. The analysis of this reaction for FCS (and FPR) measurements on open systems leads to three relaxation modes, each with an amplitude and a characteristic rate.[7] The multiple binding of ethidium bromide to DNA provides a good example of a system with a large number of coupled reactions, both for the system kinetics (T_{ij}) and for the photobleaching kinetics (P_{ij}). Experimental FPR and FCS measurements have been successfully analyzed in detail for this sytem in terms of plausible models for the kinetics of ligand binding and photobleaching.[4,5] It was necessary to carry out this analysis numerically by computer, however, because closed expressions for the eigenvalues and eigenvectors of the matrix M_{ij} were not readily obtainable in analytical form.

Analytical solutions for $G(\tau)$ and $\Delta F(t)$ are available for some simple chemical reaction systems with few components.[6,7,11] For multireaction systems in general, however, relatively extensive computations are required to obtain the full time course of fluorescence recovery or correlation decay in terms of phenomenological coefficients and assumed mechanistic kinetic models. More limited, but still useful, information can be obtained much more easily by evaluation of the initial slope of $G(\tau)$ and $\Delta F(t)$. This approach has already been proposed for conventional perturbation relaxation kinetics studies on closed systems.[12] It may easily be shown that for systems in which diffusion and chemical reactions are occurring:[13]

$$[dG(\tau)/d\tau]_{\tau=0} = \sum_{i,j} [Q_i Q_j \bar{c}_j/(4\pi^2)] \int \Phi(\nu)[-\nu^2 D_j \delta_{ij} + T_{ij}]d^2\nu \tag{10}$$

The initial slopes of $G(\tau)$ and $\Delta F(t)$ are most readily compared when the FPR measurements are supposed to performed in the limit of low extents of photobleaching. Then, from Equation 9:

$$\Delta c_j(\mathbf{r}) = I'(\mathbf{r})T \sum_{\ell} P_{j\ell}\bar{c}_{\ell} \tag{11}$$

where T is the duration of the photobleaching pulse and we have assumed that the system components were present at their equilibrium values prior to photobleaching. Let $L_j = \sum_{\ell}P_{j\ell}c_{\ell}$. Then $\Delta cj(\mathbf{r},0) = I'(\mathbf{r})TL_j$, and:

$$[d\Delta F(t)/dt]_{t=0} = [BT/(4\pi^2)] \sum_{i,j} Q_i L_j \int \Phi(\nu)[-\nu^2 D_j \delta_{ij} + T_{ij}]d^2\nu \tag{12}$$

If the photobleaching reaction for each component is independent of the others, $L_j \rightarrow \gamma_j\bar{c}_j$ and so $[d\Delta F(t)/dt]_{t=0}$ becomes identical to $[dG(\tau)/d\tau]_{\tau=0}$ if $BT\gamma_j = Q_j$. This same proportionality is also readily demonstrated for the full time course of $G\tau$ and $\Delta F(t)$.[4,6]

In the simplest and most frequently used "spot" approach a single circularly symmetric region of the system is illuminated by a beam with a Gaussian intensity profile: $I(r) = I_0$

$\exp(-2r^2/w^2)$ where w is the $\exp(-2)$ radius of the intensity distribution and I_0 is the (maximum) intensity at the center of the monitoring beam. Moreover, $I'(r) = BI(r)$ as indicated above. Then $\phi(\nu) = (P^2/4\pi^2)\exp[-(\nu w)^2/4]$ where $P = \int I(r)d^2r = \pi w^2 I_0/2$ is the laser power (monitoring level intensity) illuminating the sample area. For FPR the functions $\psi_\ell(\nu)$ depend on the kinetics and mechanism of the bleaching process. The simple assumption of independent irreversible first-order photolysis reactions for each component, (cf. Equation 8), then yields;

$$\psi_\ell(\nu) = -[P^2\bar{c}_\ell/(4\pi^2 I_0)] \sum_n \{[(-\kappa_\ell)^n/n!] \exp[-(w\nu)^2(n + 1)/8n]\} \tag{13}$$

where $\kappa_\ell = \gamma_\ell BTI_0$. Using these expressions, the characteristic time courses for correlation decay and fluorescence recovery for a single component are as follows:[7,9,10]

$$G(\tau) \propto [1 + \tau/\tau_D]^{-1} \tag{14A}$$

and

$$\Delta F(t) \propto \sum_n [(-\kappa)^n/n!][1 + n(1 + 2t/\tau_D)]^{-1} \tag{14B}$$

for diffusion, where $\tau_D = w^2/4D$.

$$G(\tau) \propto \exp[-(\tau/\tau_f)^2] \tag{14C}$$

and

$$\Delta F(t) \propto \sum_n [(-\kappa)^n/(n + 1)!] \exp\{[-2n/(n + 1)](t/\tau_f)^2\} \tag{14D}$$

for uniform flow or drift, where $\tau_f = w/V$;

$$G(\tau) \propto \exp[-(\tau/\tau_f)^2/(1 + \tau/\tau_D)]/[1 + \tau/\tau_D] \tag{14E}$$

and

$$\Delta f(t) \propto \sum_n (-\kappa)^n \exp\{-2n(t/\tau_f)^2/[1 + n(1 + 2t/\tau_D)]\} \{n![1 + n(1 + 2t/\tau_D)]\}^{-1} \tag{14F}$$

for simultaneous drift or flow and diffusion.

As an illustrative example we consider a monomolecular isomerization:

$$A \underset{K_b}{\overset{K_f}{\rightleftharpoons}} B$$

in which for simplicity we suppose that A and B have the same diffusion coefficient, D, and the same drift velocity, V. The forward and reverse rate constants are k_f and k_b. Then the equilibrium constant is $K = k_f/k_b = \bar{c}A/\bar{c}B$, and $R = k_f + k_b$ is the conventional chemical relaxation rate. Then, as has previously been demonstrated:[11]

$$G(\tau) = \{P^2/[4w^2(1 + K)]\} \{\exp[-(\tau/\tau_f)^2/(1 + \tau/\tau_D)]/[1 + \tau/\tau_D]\}$$

$$\{Q_A^2\bar{c}_A[1 + K\exp(-R\tau)] + Q_B^2\bar{c}_B[K + \exp(-R\tau)]$$

$$+ 2Q_AQ_B\bar{c}_B[1 - \exp(-R\tau)]\} \tag{15A}$$

and

$$\Delta F(t) = \{P^2/[2\pi^2 I_0(1 + K)]\} \sum_n \{(Q_A + KQ_B)[\bar{c}_A(-\kappa_A)^n + \bar{c}_B(-\kappa_B)^n]/n!$$

$$+ (Q_A - Q_B)[K\bar{c}_A(-\kappa_A)^n - \bar{c}_B(-\kappa_B)^n] \exp(-Rt)/n!\}$$

$$\cdot \exp[-2n(t/\tau_f)^2/(1 + n(1 + 2t/\tau_D))]/[1 + n(1 + 2t/\tau_D)] \tag{15B}$$

In these equations Q_A and Q_B account for the absorption coefficients and fluorescent quantum yields and κ_A and κ_b account for the extent of photobleaching of A and B, respectively. It may also be useful to note that:[13]

$$[d\Delta F(t)/dt]_{t=0} = (P^2TB/\pi w^2) \sum_{i,j} Q_iL_j(-\delta_{ij}/\tau_{Dj} + T_{ij}) \tag{16A}$$

and

$$[dG(\tau)/d\tau]_{\tau=0} = (P^2/\pi w^2) \sum_{i,j} Q_iQ_j\bar{c}_j(-\delta_{ij}/\tau_{Dj} + T_{ij}) \tag{16B}$$

III. DESIGN OF EXPERIMENTS

A. Experimental Apparatus

FPR measurements can be carried out in several different ways. In many respects the simplest approach, as discussed above, is to photobleach a small circularly symmetrical spot in the sample and then record the fluorescence from this spot as a function of time. Computations to interpret the measurements require an integration of the Fourier components of the concentrations over spatial and Fourier coordinates as in Equations 6 and 7. Another approach is based on bleaching a pattern in the sample. If this pattern is periodic, the experimental measurements can yield the Fourier components of the concentration directly and thereby simplify the interpretation of data.[14-17] The integrations over ν in Equations 6 and 7 become unnecessary. Periodic patterns are also useful for detecting anisotropic transport.[15]

Pattern photobleaching can be combined with temporal modulation in the detection process to enhance precision.[18,19] A disadvantage of both static and modulated approaches is that the pattern covers a relatively large area of the sample, which must be assumed to behave homogeneously. Hence, this approach sacrifices some degree of spatial resolution. FPR can also be carried out using video images rather than a photomultiplier to observe the fluorescence recovery.[20-22] After the spot is photobleached the entire sample is illuminated using either the laser with an expanded beam or the conventional microscope arc source. The video image is digitized to provide quantitative intensity measurements at each point of the sample. Hence, the measurement provides $\Delta f(r,t)$. This is very useful for characterizing systematic flow or drift motions in the sample. For example, using Gaussian spot photobleaching we have from Equation 5C:

$$\Delta f(\mathbf{r}, t) = QI(\mathbf{r})\bar{c} \sum_n [(-\kappa)^n/n!] \exp\{-2n[(x - Vt)^2$$

$$+ y^2]/w^2[1 + 2nt/\tau_D]\} \cdot [1 + 2nt/\tau_D]^{-1}$$

The design and assembly of apparatus for the simplest spot photobleaching measurement has been described in detail and will be discussed only briefly here.[23] Typically, an argon or krypton ion laser is used as a light source. These lasers provide steady light at high intensity and at frequencies which are appropriate for commonly used fluorophores. For rapid photobleaching, milliwatts of power on a small spot in the sample are required. Hence a 2-W laser is more than adequate for typical applications. In most FPR systems, especially those intended for measurements on cells, the optical system for focusing the excitation light onto the sample and collecting the emitted fluorescence and conveying it to a detector is supplied by an epifluorescence microscope. Auxilliary optical components are needed to convey the laser beam with an appropriate diameter into the microscope. Part of this auxilliary optical system must also modulate the excitation intensity to provide a brief intense pulse of light for photobleaching coaxial with a much less intense (typically 10^{-3}-to 10^{-4}-fold) monitoring beam. Commonly, a beam-splitting arrangement is used to divide the laser beam into a strong and a weak component.[24] The strong component is blocked by an electronically controlled shutter; the weak component impinges on the sample throughout the experiment. When the shutter is open, a second beam splitter recombines the strong and weak components into a single coaxial beam. Hence, at the beginning of a measurement the shutter is closed and the fluorescence excited by the weak monitoring beam is recorded. Then, for a brief interval, the shutter is opened to allow the bleaching beam to strike the sample. During this interval the detector must usually be protected from the intense fluorescence excited by the bleaching beam. Then the shutter closes to terminate the bleach interval, and recovery of fluorescence is monitored using the weak beam. Another elegant approach uses an acoustooptic modulator to provide high intensity and low intensity beams.[25]

In many biological specimens the level of fluorescence is very low because of the low concentration of the labeled component. Hence, FPR studies on cells usually employ a sensitive photomultiplier operating in a photon counting mode. When the measurement uses video imaging it is important to have an intensification video camera that will allow the use of very low levels of excitation intensity. This is essential to avoid damaging the specimen and photobleaching the fluorophores during the monitoring phase of the experiment.

Various functions of the experiment must be controlled electronically.[23] These include:

1. Initiating and regulating the duration of the bleaching pulse
2. Protection of the detector
3. Amplification, discrimination, and reshaping of the pulses generated in the photomultiplier
4. Counting, summing, and recording the photon counts over an appropriate series of time intervals to provide the fluorescence recovery transient.

A video frame buffer and digitizer is required for imaging experiments.

Recovery transients are conveniently analyzed with a computer. One straightforward approach fits the measured data to the appropriate theoretical curve by the use of a Marquardt algorithm.[23] An FPR measurement of diffusion yields two parameters. One is τ_D, the characteristic time for recovery, from which can be derived the effective diffusion coefficient: $\tau_D = w^2/4D$. The other is R_f, the fraction of mobile fluorophores: $R_f = [F(\infty) - F(0)]/[F(t<0) - F(0)]$ where $F(t<0)$, $F(0)$, and $F(\infty)$ are the values of $F(t)$ prior to the bleach, immediately after the bleach, and long after the bleach, respectively. Hence, $1 - R_f$ is the fraction of the bleached fluorescence that does not recover. It is taken to represent the fraction of fluorophores which are immobile on the time scale of the experiment.

In most respects, the apparatus which is used for FPR measurements is also suitable for FCS.[2,3] The electronic and optical components used for generating the bleaching pulse are, of course, unnecessary. Capability for computing the fluorescence fluctuation autocorrelation function is, however, essential. Specialized devices are available, and it is also possible to use a general-purpose computer to perform this function.[3] Because of the long duration of FCS measurements and the small size of spontaneous concentration fluctuations, the requirements for mechanical stability of the experimental apparatus and its isolation from vibration may be more stringent than for FPR measurements.

B. System Response Characteristics

1. Characteristic Times

For FCS and FPR measurements the characteristic times for transport are determined by a characteristic spatial dimension. As indicated above, for a circularly symmetric illumination intensity profile with Gaussian radius w the characteristic times for diffusion and drift are $\tau_D = w^2/4D$ and $\tau_f = w/V$. [Note that for photobleaching measurements these times are not operationally equivalent to the halftime for diffusional relaxation and the $\exp(-2)$ time for drift recovery, as they are in FCS measurements. In FPR measurements the ratios of τ_D and τ_f to these different halftimes are given respectively by the quantities γ_D and γ_f, quantities which depend on the extent of bleaching.[10]] For periodic pattern photobleaching the characteristic times are set by the pattern periodicity.[6,14-16] Chemical relaxation times, however, are independent of the spatial dimensions. For example, the chemical relaxation time for a unimolecular isomerization is $\tau_{chem} = [k_f + k_b]^{-1}$. Similarly for a simple biomolecular association such as $A + B \rightarrow C$, $\tau_{chem} = [k_f(\bar{c}_A + \bar{c}_B) + k_b]^{-1}$.[7] Hence, the measured rates for transport and chemical relaxation can be differentially varied by varying the characteristic dimension. For example, by increasing w the contributions due to diffusion and drift would be slowed while those due to chemical relaxation would be unaffected. Furthermore, transport rates and chemical relaxation rates in general depend differently on the concentrations of system components. For second and higher-order reactions the chemical relaxation times depend directly on the concentrations of the reactants as illustrated above. Effects of concentrations on transport rates will be more indirect. For example, the apparent diffusion coefficient of a molecule will decrease to the extent that it interacts with more slowly moving or immobile components. Similarly, if chemical reaction changed the charge on a molecule, this would affect its apparent electrophoretic drift velocity. Hence, varying the concentrations of system components can also be helpful in analyzing the contributions of different processes to the overall observed response.[4,5,26]

2. Amplitudes

The amplitude of the FCS correlation function, G(0), depends only on the equilibrium concentrations of the fluorescent components. It is simplest to consider the fluctuation amplitude in terms of the Poisson distribution, which governs the statistics of particle occupancy in the observation region. Assuming ideal behavior, if the average number of particles of a certain type in the observation region be n, then the root mean square fluctuation is $[<(\delta n)^2>]^{1/2} = [<(n - \bar{n})^2>]^{1/2} = \bar{n}^{1/2}$. Hence, the relative fluctuation is $[<(\delta n^2>]^{1/2}$ $\bar{n} = \bar{n}^{-1/2}$. Thus, as expected, the relative fluctuation becomes larger as the number of particles in the sampled region decreases. Therefore, fluctuation experiments are usually performed on systems with the observable components at low concentration to maximize the observable signal. In an FCS experiment using a Gaussian illumination profile the observation region does not have sharp boundaries. For a single component system: $[G(0)]^{1/2}/<i> = [\pi w^2 \bar{c}]^{-1/2}$ where $<i>$ is the mean photocount current.[7] Hence, the relative signal is as expected for the number of particles in a sample area of πw^2. More generally, for a multicomponent system:[7]

$$[G(0)]^{1/2}/<i> = \left[\sum_j Q_j^2 \bar{c}_j\right]^{1/2} / \left[(w\sqrt{\pi}) \sum_j Q_j \bar{c}_j\right]$$

In designing FCS experiments it is always useful and in many cases essential to compute the expected relative amplitude to obtain a very helpful criterion by which to assess the feasibility of the proposed measurement.

The amplitude of the FPR recovery is governed by the extent of photobleaching. The simplest photochemical mechanism supposes irreversible first-order elimination of each component, independent of all other components. From Equation 8:

$$\Delta c_i(\mathbf{r}, 0) = \bar{c}_i[1 - \exp(-\gamma_i BI(\mathbf{r})T)]$$

$$= \bar{c}_i \sum_{n=1} [(-\kappa_i)^n/n!] \exp(-2nr^2/w^2)$$

where $\kappa_i = \gamma_i BI(0)T$ and T is the duration of the photobleaching pulse. Then the fractional recovery signal is $\Delta f(\mathbf{r},0)/f = \Sigma_i Q_i \Delta c_i(\mathbf{r},0)/\Sigma_i Q_i \bar{c}_i$. The photochemistry of fluorophore labels attached to proteins, especially in the complex milieu of a cell, is poorly characterized. The assumption of independence of the photobleaching reactions is likely to be incorrect when a single component of the system carries more than one fluorophore, as in a polymerization of labeled monomers or in the binding of several fluorescent ligands to a multivalent substrate.[4-6] In these instances, the coupling of the photobleaching reactions can sometimes be accounted for in straightforward, although perhaps laborious, fashion.[4]

Amplitudes of FCS and FPR measurements provide different kinds of information. In effect, G(0) provides a way of counting the number of independent fluorescent units in the sample. If these units each contain several fluorophores and if the total number of fluorophores in the sample region is known, then measurement of G(0) can yield an estimate of the number of fluorophores per unit or aggregate. A method for determining the molecular weights of large DNA molecules based on this idea was one of the first applications of fluorescence fluctuation measurements.[27] The approach has been recently extended to study the extent of aggregation of labeled molecules on cell surfaces.[28] In this approach the cell is translated beneath the microscope objective so that the laser beam scans its surface and excites fluorescence from many sample areas. Thus, this approach, called "scanning FCS" or s-FCS, is based on fluctuations of fluorescence among surface regions rather than temporal fluctuations of fluorescence in the same region (as in conventional FCS). In FPR measurements the maximum (100%) recovery amplitude is determined by the extent of photobleaching and so provides no new information. Frequently in complex systems, however, substantially less than 100% recovery is observed ($R_f < 1$). This is usually attributed to a fraction of fluorophores which are supposed to be laterally immobile on the time scale of the FPR experiment. The immobile fraction therefore indicates the extent to which the fluorophores participate in interactions which immobilize them for periods long compared to the duration of the FPR measurement. In several applications of FPR to biological systems the measured value of the immobile fraction could be used to characterize relatively well-defined membrane interactions and so was of greater interest than the measured rates of diffusion of the mobile molecules.[29-31]

3. Acceleration of FCS Measurements

As pointed out above, a substantial disadvantage of FCS measurements arises from the long time required to record the large number of fluctuation transients which are necessary for an accurate determination of the phenomenological coefficients. One way of shortening the duration of the measurement is by using s-FCS. Within limits the scanning rate may be set by the experimenter. The faster the scanning rate, the faster the spatial fluorescence fluctuations are acquired, and so the shorter the time for integration of G(τ). There is the further advantage that the time course of the correlation decay is known in advance; G(τ) $\propto \exp[(-\tau/\tau_f)^2]$, (Equation 14C). Hence there is a single unknown, G(0), which can be more readily obtained by fitting the measured $G(\tau)$ using the known time course. There are, however, also some disadvantages to this approach. One is that information about fluorophore mobility is sacrificed by setting uniform translation of the sample to be the primary dynamic process. Hence, in the simplest s-FCS measurements the distribution (e.g., aggregation state) of the fluorophores, supposed invariant for periods of at least $\tau_f = w/V$, is the information of principal interest. A second potential problem is that the size of cells and other specimens may be too small to permit acquisition of many independent spatial fluctuations. If the maximum linear dimension of the specimen is L, then L/w is the maximum number of spatial fluctuations that can be measured from a single linear scan. Hence, multiple-line scans may be required, and the accuracy of the determination of $G(\tau)$ may be compromised.

IV. APPLICATIONS

A. Advantageous Properties

Because of their high sensitivity to concentration change and high spatial resolution, FCS and FPR are well-suited for measurements of transport and chemical kinetics on small systems. Due to the technical difficulties in the performance of the measurements FCS has been applied in relatively few instances. FPR, however, has been extensively used to measure the rates of lateral diffusion of fluorescence labeled molecules in cells and in defined cell-free systems.[6,32-35] Although neither FCS nor FPR has been extensively used to analyze chemical kinetics, the capability of both methods for this application has been demonstrated.[4,5,7,26] FPR has a set of properties which render it especially suitable for measurements on cells and on other labile or small systems:

1. Transport and kinetics can be measured under relatively normal culture conditions. The principal perturbation of cell function results from exposure to the fluorescence excitation and bleaching radiation. A variety of experiments have demonstrated that the photochemical and thermal consequences of the photobleaching pulse do not affect the values of the lateral diffusion coefficient obtained by an FPR measurement on several different kinds of systems and by several kinds of criteria.[36-39] Nevertheless, it remains possible that the cell has been perturbed physiologically or structurally in ways not detected in the diffusion measurement. Exposure of cells to low light levels for long periods of time, as occurs in an FCS experiment, might be more damaging than the exposure to a brief intense pulse in an FPR measurement.[40] This point has not been investigated because, as discussed below, FCS is little used on living cells.
2. Lateral transport over micrometer distances is measured directly. It is not necessary to infer rates of macroscopic transport from measurements of microscopic motions (as in magnetic resonance and fluorescence polarization measurements, which must be interpreted to relate the molecular motion to the observed signals).
3. The specificity conferred by fluorescence labeling permits the observation of a component present as a very minor fraction of the overall system. Of course, fluorescence labeling can also perturb function or, for antibodies, hormones, or other ligands, binding specificity. This latter possibility must be tested for each system (e.g., Reference 41).
4. A wide range of transport and chemical relaxation rates are accessible. Diffusion coefficients can range from 10^{-6} to 10^{-12} cm^{2}sec^{-1}; chemical relaxation rates, from about 20 to about 0.01 sec^{-1}.
5. In principle, systematic drift or flow can be distinguished from diffusion.[10] In practice, this requires a substantial difference between the velocities of the two kinds of processes. Specialized adaptations of FPR have been developed to detect systematic and anisotropic transport.[15,20,24]

6. Measurements are carried out on individual cells. This permits a correlation of the fluorescence dynamics with specific cellular characteristics such as morphology, extent of labeling, etc. It is also true, however, that substantial variation from cell to cell has been observed in measured diffusion coefficients.[23] Hence, frequently it is necessary to measure many cells to obtain a statistically adequate characterization of transport coefficients.

7. Measurements can be made on different regions of the same cell. This is especially interesting in cells with regional specializations such as neurons or sperm cells.[42-44]

B. Application to Cells

The principal application of FPR to cells has been in studies of the lateral mobility of cell surface components.[32-35] It has been found that the lateral diffusion of a variety of cell membrane proteins is restricted by forces in addition to the viscosity of the membrane lipid bilayer in which they are embedded. These forces could arise from interactions with structures in the cytoplasm, in the extracellular matrix, or within the membrane. Much attention has been focused on the possibility that interactions with the cytoskeleton control the mobility and distribution of membrane proteins. In the mammalian erythrocyte the chain of interactions between the membrane protein, Band 3, and cytoskeletal spectrin via the linker protein ankyrin has been documented and the effects of these interactions on the lateral mobility of Band 3 have been demonstrated.[45-48]

The plasma membrane and cytoskeleton of nucleated animal cells are biochemically more diverse and complex. The nature of the forces which constrain membrane protein mobility on these cells is still not well-understood, but it is thought that interactions between the cytoskeleton and cell surface play an important role. It has been shown that the interactions, in addition to lipid viscosity, which retard membrane protein diffusion can be essentially eliminated by separating the membrane from the underlying cytoplasmic cortex.[49-50] Furthermore, cytoskeleton-dependent interactions which retard the mobility of some surface proteins can be triggered by cross-linking other cell surface proteins.[51,52]

There are a variety of physiological processes for which the lateral mobility of cell surface proteins may be essential.[53] There are, however, few, if any, instances in which the rate of lateral diffusion has been demonstrated to limit the rate of the process overall.[54,55]

C. Measurement of Chemical Kinetics by FCS and FPR

FCS and FPR have, for the most part, been applied to study transport in biological and nonbiological systems.[6] The potential for measuring chemical kinetics has been relatively little exploited. Both methods can be used to measure chemical kinetics, either directly by monitoring the chemical relaxations, or indirectly by observing the effects of the chemical reactions on the lateral mobilities of the system components. For direct observation it is usually desirable and in many cases essential that the net fluorescence of the system change due to the chemical reaction. Hence, the absorbance coefficients and fluorescence quantum yields, represented by the quantities Q_i, must change as a result of the chemical reaction. Equations 15, in which the term containing $\exp(-R\tau)$ vanishes if $Q_A = Q_B$, provides an illustration of this requirement. An experimental demonstration of this property is provided by FCS and FPR studies of the interaction of ethidium bromide (EB) with DNA.[4-6,26] The fluorescence quantum yield of EB increases about 20-fold when it binds to DNA. Hence, this system provides a very favorable example in which to demonstrate direct chemical relaxation behavior.

Even if the chemical reactions do not change the optical properties of the system components, the effects of the reactions on measurable transport rates can provide information about kinetics and equilibrium properties. A simple illustrative example is provided by the bimolecular reaction:

$$A + B \underset{k_b}{\overset{k_f}{\rightleftharpoons}} C$$

where k_f and k_b are the rate constants for association and dissociation.[7,17,56] We shall suppose that A is a slowly diffusing nonfluorescent component with diffusion coefficient D_A; component B diffuses more rapidly with diffusion coefficient D_B and is fluorescent. The complex C has the same diffusion coefficient as A: $D_B >> D_A = D_C$. The stronger the interaction of B with A (i.e., the larger the equilibrium constant $K = \bar{c}_C/(\bar{c}_A\bar{c}_B)$), the greater will be the retardation of its diffusion. The influence of the interaction on FCS and FPR measurements will depend on the relative values of the characteristic times for diffusion and reaction; $\tau_D = w^2/4D_B$ and $\tau_{chem} = [k_f(\bar{c}_A + \bar{c}_B) + k_b]^{-1}$. If the chemical reaction is slow compared to diffusion of B (i.e., $\tau_{chem} >> \tau_D$), then any B molecule is likely to be either free or complexed but only rarely in both states during the time for fluorescence recovery or correlation decay. Hence the B molecules will appear to be in two distinct classes: fast due to diffusion of free B with diffusion coefficient D_B and slow due to diffusion of C with diffusion coefficient D_C. For D_C sufficiently small there may be negligible motion of C during the measurement period. Hence the bound B molecules will appear to be immobile.

If, however, the chemical reaction is fast compared to diffusional recovery ($\tau_{chem} <<$ τ_B), each B will react many times with A during the measurement time. Hence all B molecules will be retarded to a comparable extent, and the measurements will detect a single mobility class with effective diffusion coefficient $D_e = f_B D_B + f_C D_C$ where f_B and f_C are the mole fractions of B and C:

$$f_C = \bar{c}_C/(\bar{c}_B + \bar{c}_C) = K\bar{c}_A/(1 + K\bar{c}_A)$$

$$f_B = 1 - f_C = \bar{c}_B/(\bar{c}_B + \bar{c}_C) = 1/(1 + K\bar{c}_A)$$

Hence, an estimate of τ_{chem} could be obtained in principle by observing the transition from the fast kinetics to the slow kinetics regime. The relative rates of the diffusion and reaction processes could be regulated by varying w (to change τ_D) or by varying c_A (to change τ_{chem}). Furthermore, in the rapid reaction limit the equilibrium constant, K, could be determined from measurements of D_e vs. c_A.

V. SUMMARY

Although FCS and FPR are similar in concept and experimental implementation, the requirement by FCS for long periods of data accumulation and high-sample stability has severely limited its range of applications. FPR, however, has been used extensively to study the lateral mobility of fluorescence-labeled molecules in living cells in culture. In certain special situations it may be possible to obtain useful, if limited, information from FCS experiments with a relatively brief period of data accumulation by using a scanning approach. The scanning method is well-adapted to obtaining information about the degree of aggregation of fluorophores.[28] The latter provides a relatively simple approach to the analysis of chemical reaction systems. Both FCS and FPR are capable of probing chemical kinetics, but neither has been used much for this purpose. Nevertheless, especially for reactions which cause big changes in the fluorescence properties of the reactants, FCS and FPR could be used to study systems inaccessible to other more conventional methods.

ACKNOWLEDGMENTS

The recent work carried out in the author's laboratory that has been cited in this chapter was supported by NIH grants GM 21661 and GM 30299. I am very grateful to R. D. Icenogle for his comparative studies of FCS and FPR applied to multiple binding systems and to D. Axelrod, S. Felder, D. E. Koppel, D. Magde, N. Petersen, J. Schlessinger, and W. W. Webb for their collaboration in various phases of our work in this area.

REFERENCES

1. **de Groot, S. R. and Mazur, P.,** *Non-Equilibrium Thermodynamics,* North-Holland, Amsterdam, 1962, chap. 7.
2. **Koppel, D. E., Axelrod, D., Schlessinger, J., Elson, E. L., and Webb, W. W.,** Dynamics of fluorescence marker concentration as a probe of mobility, *Biophys. J.,* 16, 1315, 1976.
3. **Petersen, N. O. and Elson, E. L.,** Measurement of diffusion and chemical kinetics by fluorescence photobleaching recovery and fluorescence correlation spectroscopy, in *Enzyme Structure (Methods in Enzymology),* Vol. 130, Hirs, C. H. W. and Timasheff, S. N., Eds., Academic Press, New York, 1986, 454.
4. **Icenogle, R. D. and Elson, E. L.,** Fluorescence correlation spectroscopy and photobleaching recovery of multiple binding reactions. I. Theory and FCS measurements, *Biopolymers,* 22, 1919, 1983.
5. **Icenogle, R. D. and Elson, E. L.,** Fluorescence correlation spectroscopy and photobleaching recovery of multiple binding reactions. II. FPR and FCS measurements at low and high DNA concentrations, *Biopolymers,* 22, 1949, 1983.
6. **Elson, E. L.,** Fluorescence correlation spectroscopy and photobleaching recovery, *Annu. Rev. Phys. Chem.,* 36, 379, 1985.
7. **Elson, E. L. and Magde, D.,** Fluorescence correlation spectroscopy. I. Conceptual basis and theory, *Biopolymers,* 13, 1, 1974.
8. **Elson, E. L. and Webb, W. W.,** Concentration correlation spectroscopy: a new biophysical probe based on occupation number fluctuations, *Annu. Rev. Biophys. Bioeng.,* 4, 311, 1975.
9. **Magde, D., Webb, W. W., and Elson, E. L.,** Fluorescence correlation spectroscopy. III. Uniform translation and laminar flow, *Biopolymers,* 17, 361, 1978.
10. **Axelrod, D., Koppel, D. E., Schlessinger, J., Elson, E., and Webb, W. W.,** Mobility measurements by analysis of fluorescence photobleaching recovery kinetics, *Biophys. J.,* 16, 1055, 1976.
11. **Elson, E. L.,** Membrane dynamics studied by fluorescence correlation spectroscopy and photobleaching recovery, in *Optical Methods in Cell Physiology,* Vol. 40, DeWeer, P. and Salzberg, B. M., Eds., John Wiley & Sons, New York, 1986, 367.
12. **Schwarz, G.,** Kinetic analysis by chemical relaxation methods, *Rev. Mod. Phys.,* 40, 206, 1968.
13. **Elson, E. L.,** unpublished work, 1985.
14. **Smith, B. A. and McConnell, H. M.,** Determination of molecular motion in membranes using periodic pattern photobleaching, *Proc. Natl. Acad. Sci. U.S.A.,* 75, 2759, 1978.
15. **Smith, B. A., Clark, W. R., and McConnell, H. M.,** Anisotropic molecular motion on cell surfaces, *Proc. Natl. Acad. Sci. U.S.A.,* 76, 5641, 1979.
16. **Koppel, D. E. and Sheetz, M. P.,** A localized pattern photobleaching method for the concurrent analysis of rapid and slow diffusion processes, *Biophys. J.,* 43, 175, 1983.
17. **Koppel, D. E.,** Association dynamics and lateral transport in biological membranes, *J. Supramol. Struct.,* 17, 61, 1981.
18. **Lanni, F. and Ware, B. R.,** Modulation detection of fluorescence photobleaching recovery, *Rev. Sci. Instrum.,* 53, 905, 1982.
19. **Davoust, J., Devaux, P. F., and Leger, L.,** Fringe pattern photobleaching, a new method for the measurement of transport coefficients of biological macromolecules, *EMBO J.,* 1, 1233, 1982.
20. **Kapitza, H. G., McGregor, G., and Jacobson, K. A.,** Direct measurement of lateral transport in membranes using time-resolved spatial photometry (TRSP), *Proc. Natl. Acad. Sci. U.S.A.,* 82, 4122, 1985.
21. **Felder, S.,** Mechanics and Molecular Dynamics of Fibroblast Locomotion, Ph.D. thesis, Washington University, St. Louis, 1984.
22. **Wang, Y.-L.,** Exchange of actin subunits at the leading edge of living fibroblasts: possible role of treadmilling, *J. Cell Biol.,* 101, 597, 1985.

23. **Petersen, N. O., Felder, S., and Elson, E. L.,** Measurement of lateral diffusion by fluorescence photobleaching recovery, in *Handbook of Experimental Immunology*, 4th ed., Weir, D. W., Herzenberg, L. A., and Blackwell, C. C., Eds., Blackwell Scientific, Oxford, 1985, in press.

24. **Koppel, D. E.,** Fluorescence redistribution after photobleaching. A new multipoint analysis of membrane translational dynamics, *Biophys. J.,* 28, 281, 1979.

25. **Garland, P.,** Fluorescence photobleaching recovery: control of laser intensities with an acousto-optic modulator, *Biophys. J.,* 33, 481, 1981.

26. **Magde, D., Elson, E. L., and Webb, W. W.,** Fluorescence correlation spectroscopy. II. An experimental realization, *Biopolymers,* 13, 29, 1974.

27. **Weissman, M., Schindler, H., and Feher, G.,** Determination of molecular weights by fluctuation spectroscopy: Application to DNA, *Proc. Natl. Acad. Sci. U.S.A.,* 73, 2776, 1976.

28. **Petersen, N. O.,** Diffusion and aggregation in biological membranes, *Can. J. Biochem. Cell Biol.,* 62, 1158, 1984.

29. **Reidler, J. A., Keller, P. M., Elson, E. L., and Lenard, J.,** A fluorescence photobleaching study of vesicular stomatitis virus infected BHK cells. Modulation of G protein mobility by M protein, *Biochemistry,* 20, 1345, 1981.

30. **Johnson, D. C., Schlesinger, M. J., and Elson, E. L.,** Fluorescence photobleaching recovery measurements reveal differences in the envelopment of sindbis and vesicular stomatitis viruses, *Cell,* 23, 423, 1981.

31. **Axelrod, D., Ravdin, P., Koppel, D. E., Schlessinger, J., Webb, W. W., Elson, E. L., and Podleski, T. R.,** Lateral motion of fluorescently labeled acetylcholine receptors in membranes of developing muscle fibers, *Proc. Natl. Acad. Sci. U.S.A.,* 73, 4594, 1976.

32. **Cherry, R. J.,** Rotational and lateral diffusion of membrane proteins, *Biochim. Biophys. Acta,* 559, 289, 1979.

33. **Peters, R.,** Translational diffusion in the plasma membrane of single cells as studied by fluorescence microphotolysis, *Cell Biol. Int. Rep.,* 5, 733, 1981.

34. **Edidin, M. E.,** Molecular motions and membrane organization and function, in *Membrane Structure,* Finean, J. B. and Michell, R. H., Eds., Elsevier, Amsterdam, 1981, chap. 2.

35. **McCloskey, M. and Poo, M.-m.,** Protein diffusion in cell membranes: some biological implications, *Int. Rev. Cytol.,* 47, 19, 1984.

36. **Jacobson, K., Hou, Y., and Wojcieszyn, J.,** Evidence for lack of damage during photobleaching measurements of the lateral mobility of cell surface components, *Exp. Cell Res.,* 116, 179, 1978.

37. **Wolf, D. E., Edidin, M., and Dragsten, P. R.,** Effects of bleaching light on measurements of lateral diffusion in cell membranes by the fluorescence photobleaching recovery method, *Proc. Natl. Acad. Sci. U.S.A.,* 77, 2043, 1980.

38. **Wey, C.-L., Cone, R. A., and Edidin, M. A.,** Lateral diffusion of rhodopsin in photoreceptor cells measured by fluorescence photobleaching and recovery, *Biophys. J.,* 33, 225, 1981.

39. **Koppel, D. E. and Sheetz, M. P.,** Fluorescence photobleaching does not alter the lateral mobility of erythrocyte membrane glycoproteins, *Nature (London),* 293, 159, 1981.

40. **Sheetz, M. P. and Koppel, D. E.,** Membrane damage caused by irradiation of fluorescent concanavalin A, *Proc. Natl. Acad. Sci. U.S.A.,* 76, 3314, 1979.

41. **Henis, Y. I., Hekman, M., Elson, E. L., and Helmreich, E. J. M.,** Lateral motion of β receptors in membranes of cultured liver cells, *Proc. Natl. Acad. Sci. U.S.A.,* 79, 2907, 1982.

42. **Myles, D. G., Primakoff, P., and Koppel, D. E.,** A localized surface protein of guinea pig sperm exhibits free diffusion in its domain, *J. Cell Biol.,* 98, 1905, 1984.

43. **Wolf, D. E. and Voglmayr, J. K.,** Diffusion and regionalization in membranes of maturing ram spermatozoa, *J. Cell Biol.,* 98, 1678, 1984.

44. **Angelides, K. J., Elmer, L., Loftus, D., Elson, E.,** Diffusion and regionalization of sodium channels in neuronal cells, in preparation, 1985.

45. **Bennett, V.,** The human erythrocyte as a model system for understanding membrane cytoskeletal interactions, in *Cell Membranes: Methods and Reviews,* Vol. 2, Elson, E. L., Frazier, W., and Glaser, E. L., Eds., Plenum Press, New York, 1984, 149.

46. **Fowler, V. and Bennett, V.,** Association of spectrin with its membrane attachment site restricts lateral mobility of human erythrocyte integral membrane proteins, *J. Supramol. Struct.,* 8, 215, 1978.

47. **Sheetz, M. P., Schindler, M., and Koppel, D. E.,** Lateral mobility of integral membrane proteins is increased in spherocytic erythrocytes, *Nature (London),* 285, 510, 1980.

48. **Golan, D. and Veatch, W.,** Lateral mobility of band 3 in the human erythrocyte membrane studied by fluorescence photobleaching recovery: evidence for control by cytoskeletal interactions, *Proc. Natl. Acad. Sci. U.S.A.,* 77, 2537, 1980.

49. **Tank, D. W., Wu, E.-S., and Webb, W. W.,** Enhanced molecular diffusibility in muscle membrane blebs: release of lateral constraints, *J. Cell Biol.,* 92, 207, 1982.

50. **Wu, E.-S., Tank, D. W., and Webb, W. W.,** Unconstrained lateral diffusion of concanavalin A receptors on bulbous lymphocytes, *Proc. Natl. Acad. Sci. U.S.A.,* 79, 4962, 1982.

51. **Henis, Y. I. and Elson, E. L.,** Inhibition of the mobility of mouse lymphocyte surface immunoglobulins by locally bound concanavalin A, *Proc. Natl. Acad. Sci. U.S.A.,* 78, 1072, 1981.
52. **Henis, Y. I. and Elson, E. L.,** Differences in the response of several cell types to inhibition of surface receptor mobility by local concanavalin A binding, *Exp. Cell Res.,* 136, 189, 1981.
53. **Axelrod, D.,** Lateral motion of membrane proteins and biological function, *J. Membr. Biol.,* 75, 1, 1983.
54. **Gupte, S., Wu, E.-S., Hoechli, L., Hoechli, M., Jacobson, K., Sowers, A., and Hackenbrock, C. R.,** Relationship between lateral diffusion, collision frequency, and electron transfer of mitochondrial inner membrane oxidation-reduction components, *Proc. Natl. Acad. Sci. U.S.A.,* 81, 2606, 1984.
55. **Hochman, J., Ferguson-Miller, S., and Schindler, M.,** Mobility in the mitochondrial electron transport chain, *Biochemistry,* 24, 2509, 1985.
56. **Elson, E. L. and Reidler, J. A.,** Analysis of cell surface interactions by measurements of lateral mobility, *J. Supramol. Struct.,* 12, 481, 1979.

Chapter 10

CELL SURFACE CLUSTERING AND MOBILITY OF THE LIGANDED LDL RECEPTOR MEASURED BY DIGITAL VIDEO FLUORESCENCE MICROSCOPY

David J. Gross* and Watt W. Webb

TABLE OF CONTENTS

*   The present address of Dr. Gross is Department of Biochemistry, University of Massachusetts, Amherst, MA 01003.

I. INTRODUCTION

In this paper we discuss the application of video imaging in combination with digital image (diI) analysis to the measurement of the dynamics of ligand-receptor complexes on the surfaces of cells. We focus on the human low-density lipoprotein (LDL) receptor and a fluorescent analog ligand diI-LDL. Both the counting of numbers of diI-LDL particles in small clusters on the cell surface and the tracking of motions of individual receptor molecules in diI-LDL ligand-receptor complexes (LDL-RC) and small clusters of LDL-RC will be discussed in order to illustrate the applicability of digital video techniques to the study of receptor trafficking on the cell surface.

The importance of the proper cellular-level processing of plasma LDL for cholesterol control in humans has been recognized for several years.[1-3] The regulation of serum LDL concentration is based on a complex cellular feedback system;[4] its malfunctions lead to hypercholesterolemia. In particular, those individuals who express certain mutations of the LDL receptor gene that cause deficiency in expression, binding properties, or internalization of LDL-R are afflicted by the disease condition called hypercholesterolemia, which is indicated by elevated plasma cholesterol levels and by early and debilitating heart disease. It has been shown that point mutations in the gene coding for the LDL receptor are responsible for at least some of the aberrations in receptor function recognized in cases of human familial hypercholesterolemia.[5-6] One particular individual with the initials J. D. expressed a receptor which was deficient in the ability to bind to coated pits, the cell-surface organelles that are the site of specific receptor-mediated endocytosis.[5,7] Studies of J. D.'s receptor by Brown and Goldstein have played a central role in developing present understanding of LDL receptor functions. In cells from normal individuals, LDL receptors are known to aggregate in coated pits before ligand challenge,[8] and the rapid (5 to 10 min) removal of LDL bound to its cell surface receptor is known to proceed by internalization via these coated pits.[9] The elucidation of the movements of individual liganded LDL receptors and their disposition on the surfaces of normal and mutant human cells is the subject of this work and is directed toward a more detailed description of the basic subcellular events involved in the processing of LDL by individual cells.

We employ for this study a fluorescent analog of LDL that can be observed in the optical microscope on living cells without fixation. Although optical microscopy offers lower resolution compared to electron microscopic imaging, the ability to monitor movement and clustering of individual ligands on an individual living cell makes this technique equally powerful. The LDL receptor is particularly amenable for such studies for two reasons. First, the number of LDL receptors on a given cell is countable — it is on the order of a thousand. Second, the highly fluorescent LDL analog called diI-LDL is sufficiently bright that individual diI-LDL particles can be visualized in the optical microscope.[10] This ligand in combination with quantitative video microscopy allows the measurement of the disposition and tracking of individual ligands on the cell surface.

Using the techniques of fluorescence photobleaching recovery, which are described elsewhere in this volume by Elson (Chapter 9) it has been shown in this laboratory that the mobility of diI-LDL bound to its receptor on normal human fibroblasts (line GM3348) and on J. D. mutant human fibroblasts (line GM2408A) was just sufficient to account for the movement of liganded receptors to coated pits during the processing of LDL for endocytosis.[11] Based on the electrophoresis experiments of Tank et al.,[12] it appears that the unliganded LDL receptor has a much higher mobility than the liganded receptor, at least after electrophoresis. Constraints on the mobility of the liganded receptor are also released on blebs of plasma membrane.[11] In this chapter we report that liganded receptors and/or clusters exhibit very low true diffusion coefficients, but that some such receptors exhibit directed motion or flow, leading to higher apparent mobility that may be interpreted as higher diffusion coefficients. We have evidence that some, but not all, liganded LDL receptors move in a concerted fashion in some areas of J. D. cell surfaces and that others are virtually immobile.

II. METHODS

A. Cell Labeling

The methods for the production of the fluorescent analog diI-LDL and for the staining of cells with this ligand have been described in detail elsewhere.[10] Here we briefly describe the techniques involved.

LDL is isolated from outdated human plasma by ultracentrifugation. The LDL is dialyzed and lyophilized before resuspension in hexane in the presence of diI(3)C_{16}, a carbocyanine dye first described by Sims et al.[13] The reconstituted diI-LDL is further dialyzed and assayed for protein content. Unlabeled LDL is resolubilized directly from the lyophilized preparation.

The diI-LDL thus prepared retains its characteristic 20-nm-diameter spherical shape and it is thought that the lipophilic diI molecules incorporate into the phospholipid outer monolayer of the LDL particle in which apo-B, the receptor-recognizing protein, resides.[10] Barak and Webb estimated that roughly 40 diI molecules per reconstituted LDL particle were inserted into each diI-LDL particle.[10] We present evidence that diI-LDL particles bound to three different cell types do indeed contain about 40 fluorophores per particle and that such a large concentration of fluorescent tags per labeled particle allows one to quantitatively count numbers of diI-LDL particles in fluorescence spots on cells imaged in the optical microscope.

We have studied the distributions of diI-LDL bound to the surfaces of three human cell types: GM3348 normal human fibroblasts, GM2408A (J. D.) mutant human internalization-deficient fibroblasts, and A-431 human epidermoid carcinoma cells.[14] The fibroblast lines were cultured in Dulbecco's modified Eagle's medium (DMEM) supplemented with 10% fetal calf serum while the carcinoma cells were grown in DMEM plus 10% calf serum. The cells were grown in a 10% CO_2/90% air-humidified 37°C incubator. Approximately 72 hr before diI-LDL labeling, the cells were plated onto 22 × 22 mm glass cover slips; 24 to 36 hr later the standard medium in which the cells were growing was replaced by DMEM + 10% fetal calf serum from which lipoproteins had been removed by ultracentrifugation. This treatment causes the cells to upregulate (i.e., express more) LDL receptors on the cell surface.

Staining of cells with diI-LDL is straightforward. Cells are rinsed in serum-free, phenol red-free medium 199 buffer (M199) supplemented with 10 mM HEPES, buffered to pH 7.4, and chilled to 4°C. A concentration of diI-LDL of about 12 μg protein per mℓ is presented to the cells in M199 + HEPES for 15 min at 4°C, then the cells are washed three times with M199 at 4°C before being exposed to 2 mg/mℓ BSA in Hanks' Balanced Salt Solution for 10 to 30 min at 4°C to displace the nonspecifically bound diI-LDL. Cells to be observed live are rinsed two times in M199 before moving them to the microscope. Cells to be observed fixed are rinsed twice with phosphate buffered saline (PBS) and exposed to 3.7% paraformaldehyde in PBS for 5 min at 4°C before they are rinsed in PBS and observed.

B. Instrumentation

1. Hardware

Because the instrumentation hardware and software we have developed to visualize individual LDL receptor molecules has potential for broad applicability to new brilliant fluorescent receptor labels, many conceptual details are presented, since recent hardware developments have made these techniques more widely available. Our digital video instrumentation includes a three-stage intensifier vidicon camera, model TV3M from Venus Scientific®, a Grinnell® model GMR274 image-processing system, and a DEC LSI-11/73

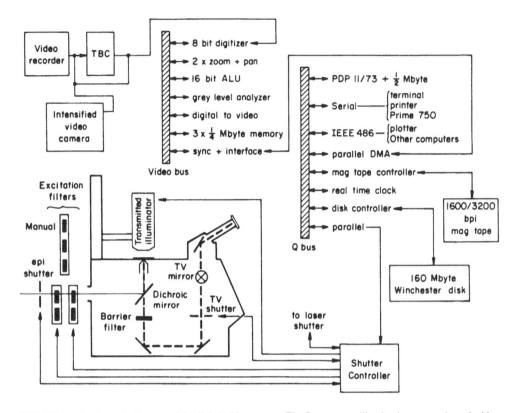

FIGURE 1.   A schematic diagram of the digital video system. The fluorescence illumination passes through either a manually-positioned interference filter or through filters positioned by computer control of two articulating arm filter holders. The positions of the articulating arms as well as the open or closed state of any of the electronic shutters in the system is under computer control, with manual override. The video camera is interfaced to the inverted microscope through a side port through which the image is passed by a sliding front-surface mirror. The video output of the camera is digitized by the Grinnell® image processor, which also performs image arithmetic, frame storage, and other image manipulations. A digital to analog converter produces RGB color video, which drives the monitor. The whole system is under computer control such that real-time acqusition, storage, and processing of time-lapse data proceeds without user intervention.

minicomputer with peripherals. A general diagram of the whole system is shown in Figure 1.

The video camera can be easily attached to either upright or inverted Zeiss® research microscopes. Much of the work described here was carried out on an adapted Zeiss Universal® microscope and more recent work has been carried out with a Zeiss® IM35 inverted microscope. As the camera produces an image at light levels below that discernible to the eye and overloads at higher light levels, the standard TV/eyepiece 60/40 splitting mirror in the IM35 transmits insufficient light for visualization when the camera is in operation. Since the optimal level of fluorescence illumination is the lowest level at which an acceptable image is produced, it is clear that maximal collection efficiency of available photons is advisable. We have thus added a sliding mirror mount which allows the insertion and removal of a front-surface 99%-reflection high-quality mirror to send all of the image light to the camera when the eyepiece is not in use.

A set of computer-interfaced electronic shutters have been added to both the Universal® and the IM35 microscopes. Both microscope epi-illumination excitation sources are gated by fast (1 msec for the Universal®, 5 msec for the IM35) shutters; in addition, the power to the condenser lamp on either microscope is computer controllable. A second shutter system has been built into the body of the IM35 microscope — the primary function of this shutter is to protect the camera or the viewer from excess light. An additional fast (1 msec) shutter is available for either microscope to control a third light source, such as a laser.

All shutters are under direct computer control via a home-built TTL logic shutter controller. The controller is interfaced to a DRV-11J parallel interface to the Q-bus. The design is such that a manually operated switch box allows the user to open and close shutters independently of the computer. Logic circuitry on the controller automatically closes the camera shutter when the auxiliary (laser) shutter is selected, either by computer control or manually by the user; monitoring of the state of each shutter is accomplished by an LED/phototransistor pair separated by an edge of one of the shutter blades. The state of the phototransistor is detected by a Schmitt trigger which monitors the voltage across a resistor in the phototransistor bias circuit.

For the IM35 epi-illuminator, a pair of solenoid-driven articulating arm filter selectors is also interfaced to the shutter controller, such that multiple excitation filter combinations can be switched in and out of the illumination pathway. The articulating arms are swung by a pair of rotary solenoids which are powered by a home-built logic-controlled power supply. The filter arms, as well as the shutters, are under software control.

The Venus® TV3M intensified camera is a three-stage first-generation intensifier vidicon camera. The pedestal (or black level) of the video output signal is set manually by the user, as are the intensifier gain and target high voltage. This ability to manually control gain and black level is crucial for the application of video imaging as a quantitative technique, since one clearly cannot quantify the output signal from an instrument for which the DC level and gain are varying in response to the input signal. In practice, one adjusts the camera for full-gain operation and minimal acceptable black level for low-light quantitative imaging.

The black level of most cameras, particularly high-gain cameras such as the TV3M, varies with position across the field of the camera. Such variation in black level is easily removed by subtraction of the image produced by the camera with no input light. As no pixel in the black level image should be below the minimum digitized video level, this camera background image effectively sets the optimal black level for the camera. In practice, one often sets this level slightly higher than ''optimum'' to allow for a small amount of drift that is inherent in many cameras.

The response function of the TV3M camera in our system is linear over a wide range of light levels, as is shown in the typical camera response curve of Figure 2. To produce this curve, the camera was exposed to a dim, but saturating, level of fluorescence light and the digitized average grey level value of a large illuminated area was measured and recorded. Then a series of neutral density filters was inserted in the excitation beam path and the average intensity over the same area was recorded. The full-scale digitized value of 255 was recorded at the brightest light level, while the camera responded linearly for lower light levels until it reached its ''black level'' value with no input light. Although the TV3M can be highly nonlinear, we have adjusted the digitizer threshold and saturation levels to match the linear response region of the camera.

For quantitative imaging, the spatial variation of gain, or shading, must also be measured and corrected in the final data-containing image. Every camera has intrinsic variation in gain; additionally, the fluorescence epi-illumination system may illuminate the sample non-uniformly. Correction for both of these variations is easily made by collecting a ''mask'' image of a highly uniform fluorescent sample with a thickness comparable to that of the objects in the data image and, after subtracting the camera black level image from the mask image and subtracting the camera black level plus background fluorescence from the data image, dividing the data image by a suitably normalized mask image. In practice, the most difficult image to collect is that of the sample background fluorescence since such background often differs in adjacent regions of a cover slip and since this background is also affected by spatial variations in gain and excitation light intensity.

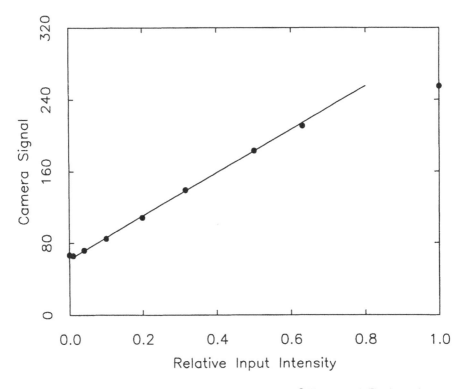

FIGURE 2.   System response function for the digitized Venus® TV3M signal. The detected camera signal is plotted vs. relative input light level. The camera was deliberately saturated at the brightest light level and the black level was set high for illustrative purposes.

We have found that 1 to 2% diI dispersed in a formvar film coated on a glass slide or cover slip[15] is a most suitable uniform thin fluorescent specimen to use for the generation of the mask image. Small imperfections in the film can be smeared uniformly in the mask image by averaging a large number of video frames while panning the microscope stage in the x- and y-directions. A uniform film of fluorescence suspended in liquid, either between a cover glass and a glass slide, or in a glass microcapillary tube is a less suitable specimen for generation of the mask image, since a large volume of fluorescence both above and below the objective focal plane is included in the image. As this region of fluorescence is not included in images of stained cells, division by such a mask image can lead to incorrect shading correction.

2. Software

We have developed a large collection of software to run the real-time image collection and image analysis hardware. A general program called VIDEO contains most of the high-level routines that control digital image storage and retrieval from the Winchester disk, digitizer operation and averaging, image grey level sampling, pseudocoloring and grey level stretching, image subtraction, ratio image generation, image shading correction, edge enhancement, and other such image operations. As far as possible, all high-level multiple-image processing is done via the pipeline arithmetic and logic unit which allows addition, subtraction, and logical operations to be performed on two 16-bit images in one or two video frame times (1/30 to 1/15 sec). Image multiplication and division is accomplished by the use of logarithmic lookup tables for input images and an exponential lookup table for the output. Such integer-based division or multiplication is very fast as compared with floating point operations on an LSI 11/73, but it lacks the full grey level resolution of an exact division. In practice, we find that the approximately 6-bit resolution of integer division of two 8-bit images is quite sufficient for application to cell fluorescence images.

A program called TIMLAP facilitates the collection of timed sequences of digital video images. The program is controlled by a real-time clock on the Q-bus for accurate image collection timing; additionally, it is synchronized with the vertical retrace of the video system. The first timing mechanism is used to allow accurate interval timing for times other than multiples of 1/60 sec. The second synchronization is necessary to illuminate consistently the video camera at the same point in its video cycle for each collected data frame. As the epi-illumination light is shut off at all times except for the few video frame times during which images are collected, this synchronization is critical to keep illumination levels constant for each image in a timed series of images, thus allowing quantitative comparisons of images as a function of time. This program is designed to allow single frame or multiple frame images to be collected for time intervals as short as about 1 sec between frames to as much as 10 min between frames for up to 999 timed intervals. Full frames or quarters of any contiguous section of a collected frame may be stored during each interval. Also, any or all of the four combinations of excitation filters may be selected for insertion in the illumination pathway and independent images collected and stored at each time point. The current frame number, as well as the time remaining until the next frame are continuously updated and displayed by the program.

A companion program called MOVIE recalls and displays as fast as possible (about 2 sec/full frame) sequences of time-lapse data to facilitate examination and measurement of changes in the images as a function of time. Other, more specialized programs facilitate the measurement of displacements of particles in a series of images, allow calculation of particle mean-square displacements, particle-particle displacement, and velocity cross correlations. Several other programs aid in the analysis of specific types of image data, such as a routine which interactively selects and reads the intensity profile of a circular image for a uniformly stained spherical cell.

Nearly all the routines previously described are written as FORTRAN subroutines which are modular and thus easily exchanged between main programs. All of these programs call low-level RATFOR subroutines which control specific, individual operations of various functions of the image processing system. Only one FORTRAN-callable assembly language subroutine is usesd to initialize the shutter controller before use. The computer operating system is RT-11.

III. EXPERIMENTAL RESULTS

A. Quantitation of diI-LDL Fluorescence: Counting LDLs on the Cell Surface

As noted in the Introduction, the disposition of LDL particles bound to the specific cell surface receptor on various types of human cells is correlated with the ability of the cell to internalize efficiently the ligand LDL. The normal LDL receptor interacts specifically with coated pits on the cell surface even in the absence of ligand so that a large proportion of LDL-R are located in coated pits. This preclustering of receptors in coated pits is quite distinctly different from the behavior of LDL-R on J. D. cells, and generally from hormone receptors such as the EGF receptor and the insulin receptor which cluster in coated pits only after ligand challenge.

The portion of the LDL-R which is responsible for the binding of the receptor to coated pits is known to occupy the region of the receptor which resides in the cytoplasm. A mutation in the receptor gene that changes the code for one tyrosine residue into that for a cysteine in this region is known to destroy the capability of the LDL-R to bind to coated pits; this mutation is the one expressed by the GM2408A (J. D.) cell line. Presumably it is responsible for the inefficient LDL internalization of the J. D. cells. Perhaps a similar mutation may affect the efficiency of internalization of LDL in the epidermoid carcinoma A-431 cells and in other internalization-deficient cells.

It is known that the fluorescent analog diI-LDL is sufficiently bright that individual particles of diI-LDL can be seen by fluorescence microscopy.[10] There are an average of about 40 diI fluorophores per diI-LDL, and their insertion in the LDL and their fluorescence are supposed to be statistically independent. Thus, the fluorescence power emitted from individual diI-LDL particles should be distributed around the average value as determined by Poisson statistics. If a large population of *individual* LDL particles, labeled randomly, with a mean number of n, noninteracting fluorophores were sampled, the expected probability distribution of measured fluorescence power per particle should be

$$P(f) = \frac{n^f}{f!} e^{-n} \tag{1}$$

where f is the number of fluorophores contributing to the fluorescence power. This Poisson distribution peaks at the value f = n; when n is 40, the distribution width $\sqrt{n}$ is about 6.3 and becomes relatively less sharp as n becomes smaller.

For a fluorescent cluster imaged in the optical microscope, the spatial structure of the spot cannot be used to determine the number of fluorescent particles located within the spot if the characteristic size and spacing of the particles is below the (diffraction-limited) resolution limit of the microscope. Such is the case with diI-LDL, since the 20 nm particles are of the order of 1/10th the minimal resolvable size and the clusters are too tight to resolve. Thus, the only way to judge the number of diI-LDL particles per cluster is to measure the fluorescence power emitted from the spot.

The fluorescence power distribution for clusters of j particles of the type containing n fluorophores, on the average, is a linear combination of single-particle Poisson distributions:

$$P_j(f) = \frac{(jn)^f}{f!} e^{-jn} = \sum_{\ell=0}^{f} P_{j-1}(\ell) P_1(f - \ell) \tag{2}$$

where $P_1(f)$ is the distribution of Equation 1. Equation 2 demonstrates that the distribution of fluorescence power for a population of particles for which n = 40 in clusters of various sizes will contain several well-resolved peaks which begin to smear at cluster numbers j when $\sqrt{jn} \approx n/2$, that is, at j $\approx$ 9 or so. In practice, measurement uncertainty reduces this resolvable limit to approximately 7.

In order to test diI-LDL as a counting marker for ligand clusters on cells, three different cell types were stained with the ligand at 4°C, a temperature which is known to inhibit the internalization of LDL in normal cells. The three cell lines employed were GM3348 normal human fibroblasts, GM2408A human internalization-deficient fibroblasts, and A-431 human internalization-deficient epidermoid carcinoma cells. Immediately after BSA competition of nonspecifically bound diI-LDL, the cells were fixed in 3.7% formaldehyde. Fluorescence images of the three different cell lines were collected and stored in the digital imaging system for later analysis.

After the images thus collected were corrected for the spatial gain variation as described above, the fluorescence power of individual fluorescent spots was measured by the use of a specialized interactive program which calculated the total fluorescence power collected in an area of pixels selected by the user. The local background from that area was subtracted, leaving the fluorescence power due to the diI-LDL in the spot. After a large number of spots were examined, the program binned the data, displayed it to the video monitor, and fitted the data by a least-squares routine to the theoretical curve expected from the distributions of Equation 2. The details of the measurements have been described elsewhere.[16,17]

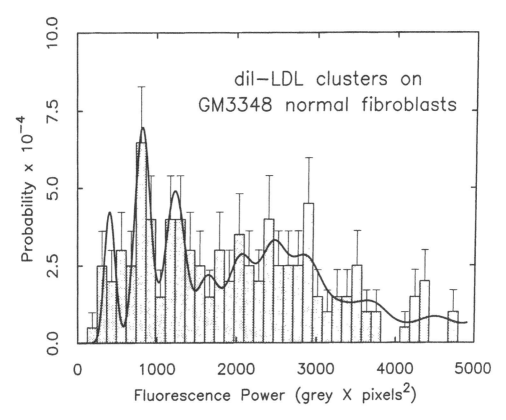

FIGURE 3. The distribution of number of spots detected as a function of fluorescence power for a GM3348 normal human fibroblast stained with diI-LDL. The data are binned over increments of fluorescence power and are normalized such that the area under the curve is unity. The bars indicate the statistical uncertainty due to number of counts per bin. The heavy smooth curve is the least-square fit to the data as described in the text.

After the data for all cells were collected, they were combined to determine the best-fit values of n, the mean number of fluorophores per diI-LDL, and the fluorescence power of a single diI molecule. Armed with these values, the data for the various cell types were examined to determine the best-fit distribution of numbers of diI-LDL particles in the spot clusters on the cell surfaces. The distribution of collected data and theoretical fit for a single GM3348 fibroblast is shown in Figure 3. The stippled bars represent the number of counts present in each bin normalized such that the integral of the distribution is unity. The error bars indicate the statistical uncertainties proportional to the square root of the number of counts in each bin. Note that peaks in the data distribution correspond to peaks in the theoretical distribution shown by the heavy solid line. As the expected probability distribution of fluorescence powers is now known, the measured value of fluorescence power for each measured spot can be compared to the theoretical distribution and thus the number of diI-LDL particles in the spot can be counted statistically. If the power of a spot is greater than that expected for a particle in the jth distribution, but less than that for (j + 1)th distribution, the value is indeterminant and thus the spot contains either j or (j + 1) particles. But if the fluorescence power of a spot lies under the jth probability peak such that its probability of residing in the (j − 1) or (j + 1)th distribution is 10% or less of it being in the jth peak, the value of j LDL particles can be reasonably assigned to that spot. By employing this criterion plus the measured locations of diI-LDL spots from which the data were taken, it is possible to construct a map of LDL clusters on the cell surface. Such a map is shown in Figure 4, along with the raw digital image from which the data of Figure 3 were taken. In

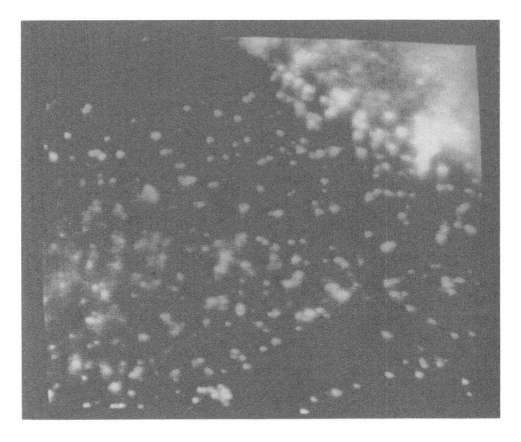

A

FIGURE 4.   (A) The digitized fluorescence image for the cell which produced the data of Figure 3. (B) A map of the numbers of diI-LDL particles contained in each fluorescent spot for which data was collected on the cell of 4A. The criteria for judging number per cluster are described in the text. A plus sign following a number indicates that the numbers in that cluster were statistically indeterminant between n and n + 1 except that 7 + denotes all spots containing 7 or more diI-LDL particles.

the cluster map the numbers refer to the number of diI-LDL per spot; indeterminate values are noted with a number and a plus sign (i.e., the value indeterminate between 3 and 4 is noted as 3 +), while all values of 7 and beyond are denoted as 7 +.

We have also analyzed data from J. D. and A-431 cells which we published elsewhere.[17] The cluster maps for these two cell types are shown in Figures 5 and 6, along with the raw fluorescence images. Comparing Figures 4, 5, and 6, it is apparent that the internalization-deficient J. D. and A-431 cells have smaller clusters of LDL bound to their surfaces than do the normal cells, a finding which is consistent with the hypothesis that LDL receptors do not cluster within coated pits on the internalization-deficient cells.

By the above analysis, both the J.D. and A-431 cell-surface LDL clusters were predominantly (>70%) of size 1 to 3, but on normal cells only about 35% of the clusters were that small.[17] Ligand challenge of normal cells at a temperature (22°C) at which internalization proceeds, leads to formation of large clusters, each containing from 19 to 32 diI-LDL particles during the early stages of internalization. These large clusters appear in the same focal plane as the surface-bound diI-LDL; thus, they represent either cell-surface clustering of ligand-receptor complexes or ligand-containing invaginations or vesicles very near the plasma membrane.

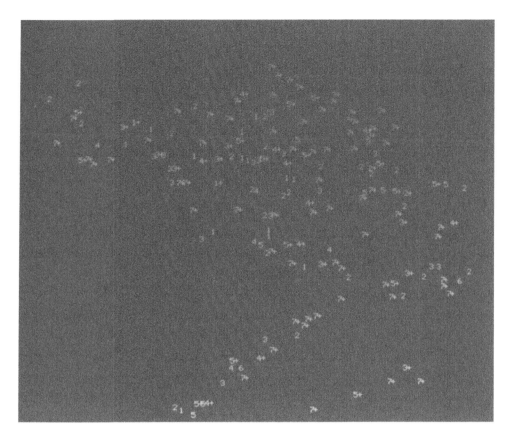

FIGURE 4B.

B. Mobility of the diI-LDL-Receptor Complex on Living Cells

1. Individual Ligand-Receptor Detection

As had been noted by Barak and Webb[10] and confirmed above, the fluorescent analog diI-LDL is bright enough that individual particles are visible in a conventional fluorescence microscope. This attribute allows one to monitor the displacements of a single particle or a small cluster of particles of diI-LDL on the surface of a living cell as the ligand-receptor complex interacts with the various constituents in its neighborhood.

Most measurements of cell-surface mobility represent the average motions of large numbers of macromolecules. The early experiments of Edidin and Frey[18] considered the gross motions of whole complements of proteins from one cell to another upon cell fusion. More quantitative techniques also sample a large number of fluorescently labeled macromolecules; one such technique is fluorescence photobleaching recovery (FPR).[19] Usually the fluorescence excited in a micron diameter area or a pattern of a few-micron periodicity on the plasma membrane is sampled to measure the time course of the recovery of fluorescence after photobleaching the same area. Results from such experiments on a cell membrane typically show a time course characteristic of diffusion of fresh fluorophore into the photobleached area. However, the fluorescence of labeled proteins usually shows incomplete recovery, indicating that a portion of the fluorescent label in the bleached area is immobile (i.e., it cannot be replaced by fresh fluorophore during the maximum duration of an experiment of tens of minutes or so). The remainder of the fluorescence recovers in accord with a characteristic diffusion coefficient which for proteins is usually a few orders of magnitude smaller than hydrodynamic theory would predict. These two parameters, the immobile fraction and the diffusion coefficient, represent an ensemble average involving

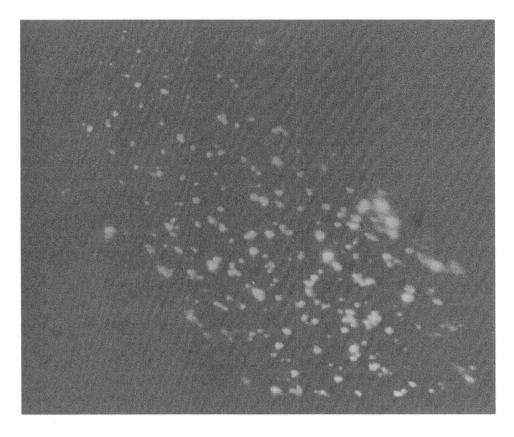

A

FIGURE 5.   (A) Digital fluorescence image of a GM2408A (J. D.) human mutant fibroblast. (B) The map of cluster sizes for this cell. (Reproduced from the *Biophysical Journal*, 1986, 49, 901—911 by copyright permission of the Biophysical Society.)

the labeled molecules in the bleached spot and many more in the surrounding diffusion field. The motion of an individual diffusing molecule in the population might deviate considerably from the mean behavior, but would not be identified in the ensemble.

By employing a pattern-exposure photobleaching method to cells stained with diI-LDL, Barak and Webb were able to measure effective diffusion coefficients of an LDL-receptor complex on GM2408A internalization-deficient fibroblasts with only about one fluorophore-labeled complex per micrometer squared. At 21°C they found D ≈ 1.4 × 10^{-11} cm^2/sec and a mobile fraction of about 75%. At 10°C the diffusible fraction of diI-LDL bound to receptors on these cells decreased to about 20% with a poorly characterized diffusion coefficient of ~0.5 − 3.0 × 10^{-11} cm^2/sec. Similar values were obtained for normal fibroblasts at 10°C. This crucial experimental point was accessible only because low temperature slowed internalization. It is quite difficult to monitor the surface mobility of a ligand-receptor complex which is undergoing internalization, as is the case for LDL on normal fibroblasts at room or higher temperature. Because the measured mobilities of bound diI-LDL are the same on both the internalization-deficient J. D. cells and on normal fibroblasts at 10°C, the measurement of mobilities on J.D. cells at higher temperatures should reflect that part of the LDL-receptor complex mobility that is not directly related to coated-pit binding and the internalization pathway.

Barak and Webb have also demonstrated that the individual mobilities of diI-LDL particles

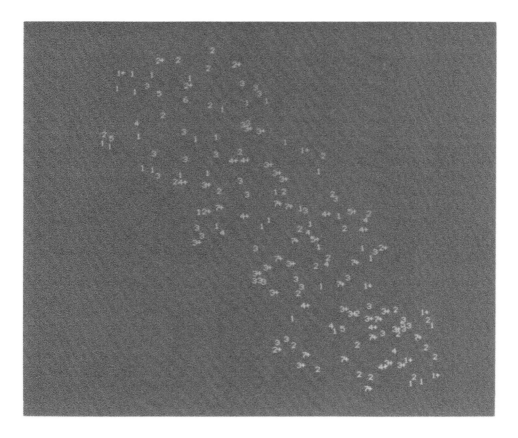

FIGURE 5B.

or clusters on blebs induced in the plasma membrane are measurable by observing particle trajectories over times of a few seconds by video fluorescence microscopy; the motion appeared to be diffusive Brownian motion with diffusion coefficients of $\approx 1 - 2 \times 10^{-9}$ cm^2/sec and no immobile fraction on the bleb.[11] Enhanced diffusibility of cell-surface proteins on bleb membrane has been reported for membrane proteins in other systems.[20] It is possible that interactions of membrane proteins with the cytoskeleton may play a role in the modulation of protein mobility in intact cell membranes, and that release of restraints on protein diffusion on blebs is mediated by separation of the membrane from the cytoskeleton during the blebbing process.

Tank et al. determined the mobility of the LDL receptor in the absence of ligand by measuring the time course of the postelectrophoresis relaxation of a segregated distribution induced by application of electric fields.[12] These postelectrophoresis diffusion coefficients appear to be enhanced by at least one order of magnitude and the internalization kinetics of the receptor also seems to have been enhanced by electrophoresis. The techniques for tracking of individual LDL-RC that we report here were developed with the objective of sorting out the puzzling behavior of the mobility and interactions of LDL-RC revealed in these earlier experiments.

2. Time Lapse Data Collection and Analysis of LDL-RC Mobilities

The hardware and software operation of the video system during timed interval data collection is described previously in the Methods section. It is most useful to employ time-lapse image recording with the diI-LDL system on intact cells for two reasons. First, the motions of diI-LDL particles bound to receptors on the cell surface tend to be so slow that

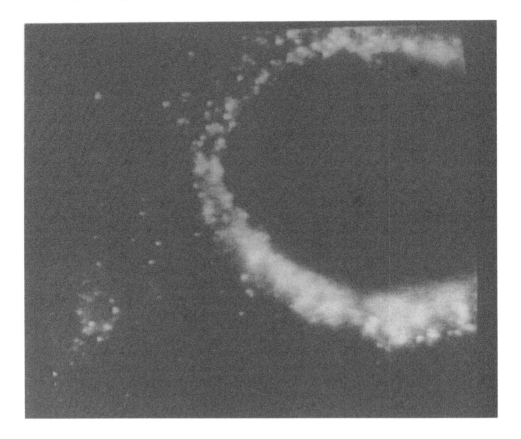

A

FIGURE 6.   (A) Digital fluorescence image of an A-431 human epidermoid carcinoma cell. (B) The map of cluster sizes for this cell. (Reproduced from the *Biophysical Journal*, 1986, 49, 901—911 by copyright permission of the Biophysical Society.)

several seconds or even minutes must elapse between frames in order to observe measurable particle displacements. Second, even though diI is resistant to photobleaching, intermittent illumination is necessary to reduce photobleaching during the 20 to 30 min needed for mobility experiments.

For most experiments, fluorescence images were collected at 20 sec intervals over 20 to 30 minutes for a total of 60 to 90 collected images. Cells were illuminated for a total of 120 msec per frame, for a total exposure time of 10.8 sec for a 30 min data run. This exposure time is negligible compared to the several minute photobleaching time at these excitation levels. Although a single-frame exposure requires only 30 msec, the slow camera faceplate response time extends the exposure required to obtain full sensitivity. The Venus® TV3M, a relatively slowly responding camera, requires on the order of 100 msec to approach full response.

Several procedures for receptor mobility measurements are described below in order of increasing effectiveness. To illustrate the capabilities of digital image analysis, we analyze a set of noisy low contrast images obtained with a demonstrator microchannel plate-intensifier camera.

Two images of a GM2408A fibroblast stained with diI-LDL are shown in Figure 7. The first frame displays the initial distribution of liganded receptors at time zero, the second shows the distribution 68 time-lapse frames later, after 1,360 sec or 23 min had elapsed.

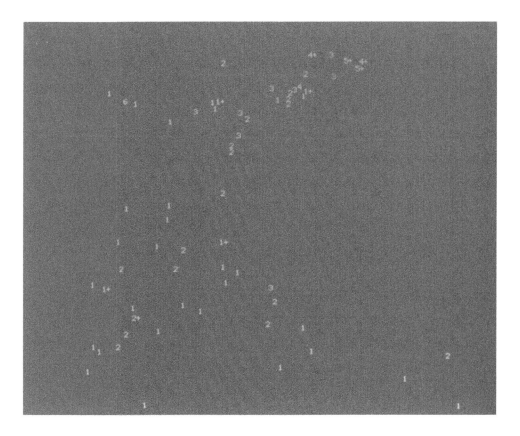

FIGURE 6B.

The distribution of ligand-receptor complexes on this cell has changed significantly over this time period, as can be seen by careful comparison of the two images. The cell fluorescence images in Figure 7 are obscured in the lower left-hand corner of both frames by a severe background-level variation, due to the camera used to collect these images. The leading edge lamellipodium of the cell is just discernible in the two images at the right edge of the cell. A large, stained protuberance on the lamellipodium corresponds to an extraneous deposit visible under phase-contrast observation.

A more direct way to examine the motion of particles in such sequential images is to isolate only the parts of the fluorescence signal which change from frame to frame. Most simply, sequential differential images can be generated by subtracting successive fluorescence frames. Such a differential image for the cell of Figure 7 is shown in Figure 8A. This image is data frame 3 minus data frame 2 displayed with the zero difference level set at a value of 128, a medium-grey intensity. Any bright feature that has moved will thus show a characteristic bright/dark edge perpendicular to the direction of motion. Particles which move more than their image diameters will appear as a pair of bright and dark spot pairs. Such motion is evident in Figure 8A. However, this method is not useful because it is difficult to detect very small changes in position or much larger changes of relatively dim features. In addition, very small chanes in the vertical position of the cell relative to the objective focal plane, whether due to optical drift or to cell movement, confuse the analysis. As focus drifts, the image of a point source spreads and loses contrast, perturbing the differential image excessively.

A reliable but tedious method to examine the motion of the diI-LDL particles bound to their cell-surface receptors is to track manually the fluorescent spots for a complete sequence

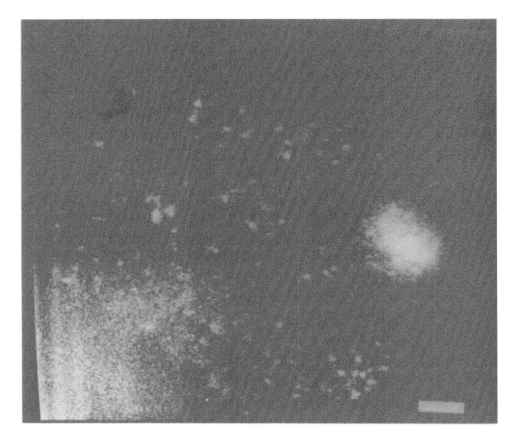

A

FIGURE 7.   Two low-contrast noisy digital fluorescence images of a GM2408A fibroblast taken during a time-lapse series. See text. (A) Frame 1 at t = 0. (B) Frame 69 at t = 23 minutes. Note changes in receptor distribution.

of frames.[21] (Barak and Webb used a crude version of this approach to measure the fast diffusibility of diI-LDL bound to receptors on plasma membrane blebs.[11]) Since our data is stored in the form of digital images, it is straightforward to apply the digital video system to collect the x and y pixel coordinates of the center of selected fluorescence spots and thus to follow a series of spots over time. A specialized program which displays all data frames sequentially allows the user to store the location of a large number of fluorescent spot locations by moving a joystick-controlled cursor to the location of a spot and typing one keystroke on the terminal. The program is designed to overlay two images: with the current image displayed in green and a previous image displayed in red so that reference points are available for fast-moving particles in complicated collections of particles. Provision is made for skipping data frames if a particular particle is lost temporarily, while refocusing, for example. A loop in the program is designed to enter fiducial drift information from either a visible mark stuck to the substrate or from a series of phase, DIC, or brightfield images collected simultaneously with the data frames. We have found it simplest to match inter-actively the entire image of each data frame pair by comparing a frame displayed in green overlaid with the first data frame in red and to measure fiducial drift from the magnitude of x,y translation necessary for superposition.

Once the measured particle displacements and fiducial drift of the cell have been recorded

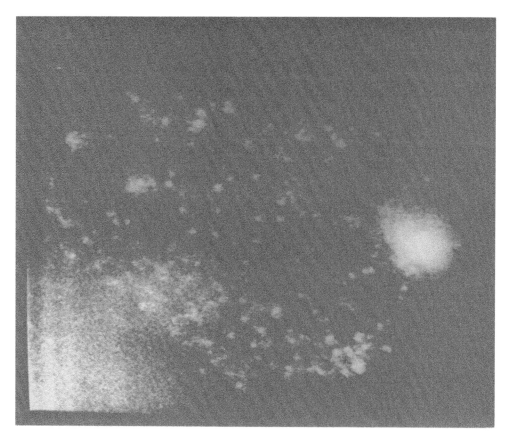

FIGURE 7B.

and collected in computer memory, it is a simple matter to display the particle motions. A plot of displacements of several of the diI-LDL spots for the cell of Figure 7 is shown in Figure 8B. The initial location of each particle is indicated by a filled circle. The movements of the particles for the 23 min of observation are shown by the line tracks. One immediate conclusion that can be drawn from this plot is that some of the particles, particularly the three at the right in the center of the lamellipodium, appear to have undergone coordinated directed motion, qualitatively different from the random motion one would expect for simple diffusion in a uniform medium.

One can quantitatively assess the motion of individual spots by calculating the particle displacements as a function of time increments between measurements. For pure diffusion the mean-square displacements $<r^2(\delta t)>$ for each value of time increment δt should plot as a straight line with a slope of 4D where D is the diffusion coefficient. For particles moving with a constant velocity v, the function $<r^2(\delta t)> = v^2(\delta t)^2$. The mean-square displacement is determined from experiment as:

$$<r^2(\delta t)> = <r^2(n\delta t_1)> = \frac{1}{(N - 1 - n)} \sum_{j=0}^{N-1-n} ([x(j\delta t_1 + n\delta t_1) - x(j\delta t_1)]^2 + [y(j\delta t_1 + n\delta t_1) - y(j\delta t_1)]^2) \qquad (3)$$

where N is the total number of measured particle locations, δt_1 is the interval between the periodic displacement measurements, and n and j are positive integers, n determining time increment $\delta t = n\delta t_1$. This summation assumes a steady-state process for the duration of the

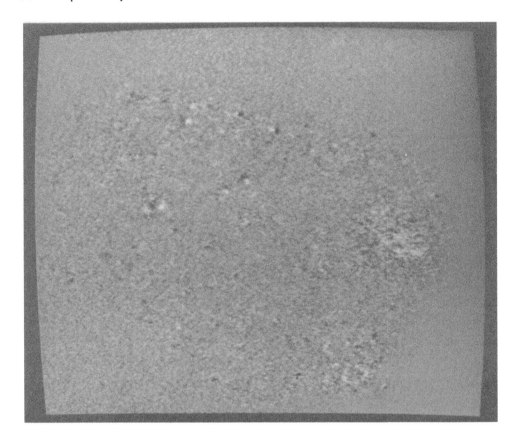

A

FIGURE 8.   (A) Difference digital image for the cell of Figure 7. The frame 2 image was subtracted from the frame 3 image with zero difference mapped to a mid grey level. Motion is barely detectable as dark/light pairs in the difference image. (B) Displacements of the diI-LDL spots of Figure 7 as tracked at 20 sec time intervals for 69 frames. The t = 0 spot locations are marked by filled circles. Note the characteristic directed motion of many of the particles. The numbers reflect only the order in which the particles were tracked.

experiment. The maximum available number of measurement intervals N' for diffusion interval $\delta t = n\delta t_1$ is $N - n$; note that the number of nonoverlapping intervals N_0 is only the integer part of $N_0 = N/n$.

The experimental uncertainty of determination of $<r^2(\delta t)>$ for pure diffusion depends on the number of independent observations and on the particle probability density $P(r)dr = [r\exp(-r^2/4D\delta t)]/(2\pi D\delta t)dr$. Thus, $<r^2(\delta t)> = 4D\delta t$ with the density of values

$$P[<r^2(\delta t)>]dr = ([\exp(-r^2/4D\delta t)]/2D\delta t)dr$$

The second moment of this density is also $\sigma[<r^2(\delta t)>] = 4D\delta t$. Thus, the standard deviation of the determination of $<r^2(\delta t)>$ for N-independent experimental intervals is approximately $\sigma[<r^2(\delta t)>] \simeq 4D\delta t/\sqrt{N}$. Therefore, the values of $<r^2(\delta t)>$ are well-determined for small δt in an experiment of finite duration, but the uncertainty increases with δt. This large uncertainty for large n is not reflected in scatter along a plot because of the interval correlations of the method. Independent trajectories would show the uncertainty in a broad distribution of fitting values of D.

Values of $<r^2(\delta t)>$ for the four particle trajectories shown in Figure 8 are plotted in

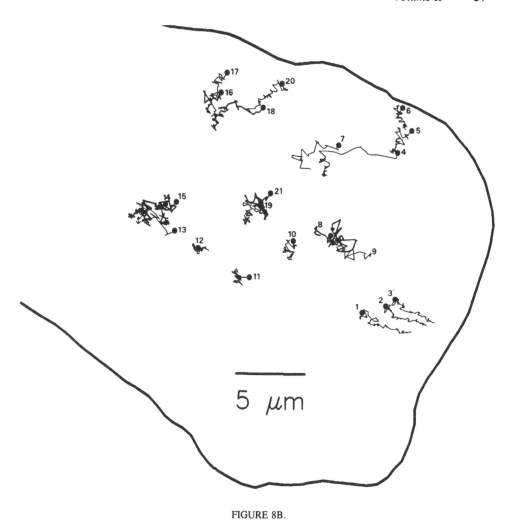

FIGURE 8B.

Figure 9 up to $\delta t = N\delta t_1/2$. Straight-line plots suggest diffusive motion for two of these particles but one is so nearly immobile ($D \gtrsim 10^{-12}$ cm^2/sec) that the residual random motion suggesting a roughly linear plot of very low slope may be nearly all attributable to position measurement uncertainty. However, the mean square displacement of the particle selected from the group of three showing nearly straight parallel trajectories in Figure 8 is parabolic, as expected for constant velocity translation.

Particles undergoing uniform directed motion at roughly constant velocity also may show superimposed random motion that can be recognized by examining the mean-square displacement of the components of motion parallel and perpendicular to the mean-directed velocity. If x' is defined as the parallel direction and y' the perpendicular direction, then:

$$\langle x'^2(\delta t) \rangle = v_x^2(\delta t)^2 + 2D_x \cdot \delta t$$

and

$$\langle y'^2(\delta t) \rangle = 2D_y \cdot \delta t \qquad (4)$$

Such a resolution of components is shown in Figure 10A for Particle 5 where the parabolic behavior is seen. A least-squares fit for these data gives $v = 2.8 \times 10^{-7}$ cm/sec,

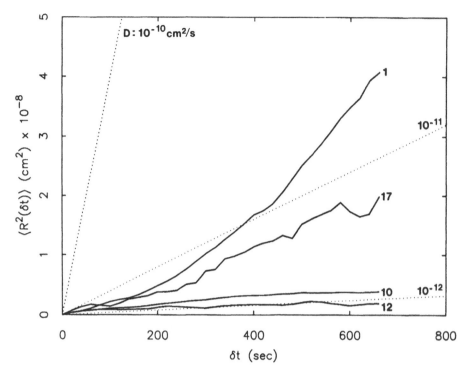

FIGURE 9. Mean-square displacements of four of the particles of Figures 7 and 8. A nonlinear least-square fit to the function $<R^2(\delta t)> = v^2(\delta t)^2 + 4D\delta t$ produced the following best-fit values of v and D: $v_1 = 5.7 \times 10^{-7}$ cm/sec, $D_1 = 4.2 \times 10^{-12}$ cm²/sec; $v_{17} = 2.8 \times 10^{-7}$ cm/sec, $D_{17} = 4.5 \times 10^{-12}$ cm²/sec; $v_{10} = 0$ (not significant), $D_{10} = 2.0 \times 10^{-12}$ cm²/sec, $v_{12} = 0$ (not significant), $D_{12} = 1.2 \times 10^{-12}$ cm²/sec.

$D_{x'} = 6.4 \times 10^{-13}$ cm²/sec and $D_{y'} = 5.4 \times 10^{-13}$ cm²/sec. $D_{y'}$ probably and $D_{x'}$ perhaps reflect more measurement uncertainty than particle diffusion. This particle may have been secured to a macroscopic moving structure, perhaps a component of the cytoskeleton.

The same x' and y' breakdown was attempted for Particle 1 shown in Figure 10B. Only an incomplete breakdown into components could be accomplished for this particle because its directed displacement follows a curvilinear rather than a straight-line path, as can be noted in Figure 8. Although the statistical uncertainties for large δt are excessive, this particle like many others (not shown) tends to suggest a concave upward plot of $<(r)^2(\delta t)>$. This behavior has been found for over half the particles on nearly all of the 20 or so GM2408A cells we have examined. This behavior implies inhomogeneous directed motion that remains to be described in future reports.

At 8°C ligand-receptor complex motion was very slow or not detectable over the 30 min of data-frame collection. GM2408A cells examined at 37°C showed increased ligand-receptor mobility. This temperature dependence and the appearance of an immobile fraction of labeled receptors agrees generally with the FPR results of Barak and Webb.[11] However, a closer comparison suggests smaller values of D for our tracking experiments than for the FPR data; additionally, directed motion was not detected by Barak and Webb. It has been shown that the characteristic shapes of photobleaching recovery curves are quite different for a population of fluorophores flowing at constant velocity rather than diffusing.[19] However, this result holds only for locally homogeneous flow over the area replenishing the photobleached region. If these vector displacements are randomized over distances comparable to the scale of interest, which is the bleaching pattern scale of a few microns, then an ensemble average measurement of random directed displacements is not distinguishable from molecular dif-

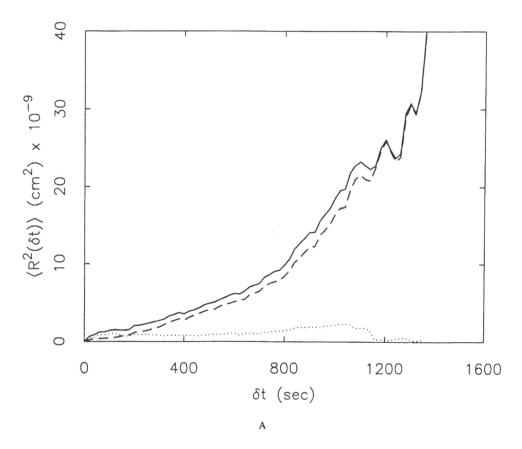

A

FIGURE 10. Mean-square displacements resolved into the directions parallel (dashed) and perpendicular (dotted) to the mean direction of motion. The solid line is the total mean-square displacement. (A) Particle 5. (B) Particle 1.

fusion. Therefore, in FPR experiments there is no separation of diffusive motion from a random pattern of short range tractions and a larger effective diffusion coefficient is inferred from experiment. Such locally uniform but globally random-driven motion leads directly to an equation of motion which is comparable to diffusion. For the above model with particles moving with uniform speed with randomly directed orientation the flux density of particles on a plane is

$$\vec{J}_v = - \frac{|\vec{v}|\ell}{2\sqrt{2}} \nabla C \tag{5}$$

where $|\vec{v}|$ is the mean speed of the particles and ℓ is the characteristic path length over which the particles move without changing direction. The effective "diffusion" coefficient D_v for randomly directed particle motion is $|\vec{v}|\ell/2\sqrt{2}$. Our moving liganded LDL receptors have averaged speeds of roughly $|\vec{v}| \approx 2 \times 10^{-7}$ cm/sec. With inferred characteristic distances of $\ell \approx 5$ μm, these values yield an effective "diffusion", coefficients $D_v \approx 3.5 \times 10^{-11}$ cm²/sec at room temperature, in excellent agreement with the value of $D = 1.4 \times 10^{-11}$ cm²/sec measured by Barak and Webb. Thus, short-range driven molecular motion, such as that imagined for small-scale contractile cytoskeletal attachments could appear to be indistinguishable from diffusive motion. Work in progress aims to explore the occurrence of this effect on the cell surface. This mechanism is of course complicated by the expected variation of $|\vec{v}|$ with ℓ as in turbulent diffusion.

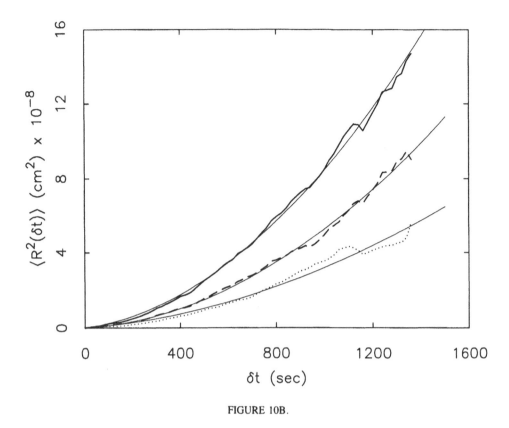

FIGURE 10B.

Eisinger et al.[22] have suggested that the directed LDL-receptor complex motion we detect is the result of directed diffusion driven by concentration gradients associated with inhomogeneous cell-surface protein distributions. Thus, particles moving by true diffusion processes would appear to undergo directed flow, due to flux generated by preestablished concentration gradients. Decay of preexisting concentration gradients would lead to nonstationary conditions detectable in the data.

A crude analysis for nonstationary processes is accomplished by dividing single-particle data sets into sequential subsets and analyzing the data as independent subsets. Figure 11 demonstrates two such analyses for Particles 1 and 2 of Figures 7 and 8. The heavy, solid curves on both graphs show the mean square displacements for the full set of particle locations, while the lighter dotted curves are the mean square displacements for the first half and the second half of the displacement data of these particles; the second half of the data is displaced on the δt axis for clarity. These data suggest that nonstationary processes are involved in the motions of LDL-RC. Their trend has the opposite sense of Eisinger's effect. However, these results are marginally significant, due to statistical uncertainty. Although active nonstationary LDL-RC mobility is implied by the data, one must be cautious, as degradation of the cells after long times of observation may contribute to this effect.

Pharmacological studies indicate that some of the constraint on receptor mobility can be released. Preliminary illustrative data are shown in Figure 12. Figure 12A shows a GM2408A cell incubated with 30 μM trifluoperazine (TFP) in M199 + HEPES for 5 min at 4°C, then stained and mounted as described in Methods, with the exception that all solutions contained 30 μM TFP. Figure 12B is the fluorescence difference image of the TFP-treated cell with 15 sec between frames. Clearly, this cell exhibited a marked increase in the mobility of individual ligand-receptor complexes as compared to the untreated cell of Figure 8A. It showed large global motion of all fluorescent spots instead of only a few spots in the untreated

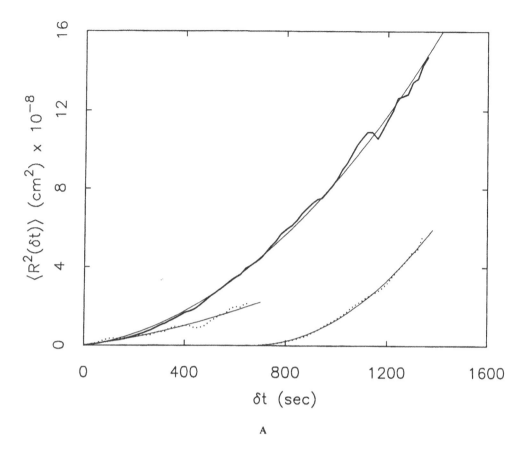

FIGURE 11. Total mean-square displacement for the first and last half of a particle track for Particles 1 (A) and 2 (B). The first half of the data was analyzed and plotted independent of the second half. For clarity, the origin of the second half of data was shifted halfway along the δt axis. The total data set mean-square displacement is shown by the solid line, the two halves by dotted lines. Best fits to the function $<r^2(\delta t)> = 4D\delta t + v^2(\delta t)^2$ are shown by the light solid lines. For Particle 1, $v_{tot} = 5.2 \times 10^{-7}$ cm/sec, $D_{tot} = 4.2 \times 10^{-12}$ cm^2/sec, $v_{early} = 3.0 \times 10^{-7}$ cm/sec, $D_{early} = 4.2 \times 10^{-12}$ cm^2/sec, $v_{late} = 6.8 \times 10^{-7}$ cm/sec and $D_{late} = 1.3 \times 10^{-12}$ cm^2/sec. For Particle 2, $v_{tot} = 4.2 \times 10^{-7}$ cm/sec, $D_{tot} = 4.8 \times 10^{-12}$ cm^2/sec, $v_{early} = 2.4 \times 10^{-7}$ cm/sec, $D_{early} = 3.7 \times 10^{-12}$ cm^2/sec, $v_{late} = 5.4 \times 10^{-7}$ cm/sec and $D_{late} = 1.4 \times 10^{-12}$ cm^2/sec.

cell. Preliminary analysis of these data indicates that the mobility of diI-LDL on the cell surface has been enhanced in the presence of TFP by somewhat more than an order of magnitude. Nevertheless, some evidence for preservation of directed motion appeared and is being studied by R. Ghosh in our laboratories.[27]

We have also found that 30 μ*M* TFP inhibits the internalization of diI-LDL by *normal* GM3348 fibroblasts.[23] Surface-bound diI-LDL on *normal*, TFP-treated fibroblasts moves about on the cell surface at room temperature, although it may still be clustered in coated pits.[28] Since Tank et al. showed that coated pits are not electrodiffusible on untreated fibroblasts,[12] we suppose that they do not move on the cell surface. Perhaps TFP releases constraints linking at least some cell-surface structures with cytoskeletal elements. TFP is thought to act semispecifically on calmodulin, a Ca^{+2}-regulatory protein that modulates actin-myosin interaction. We have shown that TFP does disrupt the F-actin cytoskeleton of normal fibroblasts.[23] These results lead us to infer that cytoskeletal dynamics plays an important role in the motion of cell-surface bound diI-LDL receptor complexes.

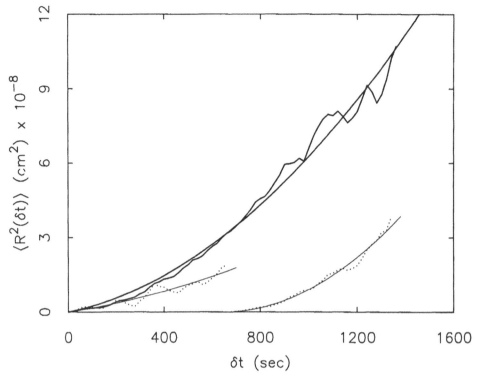

FIGURE 11B.

IV. DISCUSSION

We have demonstrated that the combination of digital video image processing and analysis with an intensely fluorescent analog of LDL allows the detection, identification, and tracking of individual LDL particles bound to the cell surface and counting of the numbers in clusters. Tracking of the motion of these individual particles and small clusters of the LDL-receptor complex is beginning to provide new insight into the mechanisms of their motion on the cell surface. We have shown that individual diI-LDL particles are detectable, that they contain about 40 fluorophores per LDL particle, and that the distribution of cluster sizes corresponds well with expectations of receptor binding to coated pits and the ability of the cell to endocytose LDL through coated pits. We have found that the motion of LDL-receptor complexes sometimes includes a dominant linear component of the motion which suggests that contractile cytoskeleton plays a role in LDL-receptor movement.

These techniques provide a powerful new tool for the study of the cell-surface trafficking of membrane macromolecules that can be labeled with sufficiently bright fluorescence. The availability of brightly fluorescent phycobiliproteins[24] which contain unitary fluorescence equivalent to about 35 fluorescein molecules offers the possibility of applying fluorescent labels that allow the tracking and counting of receptors by the methods described here. Since the phycobiliprotein fluorescence is generated by a specific fluorophore structure instead of a statistical assembly of fluorophores, as in diI-LDL, the statistical spread of fluorescence is avoided. Presuming that one-to-one coupling of a phycobiliprotein to a receptor can be achieved, the detection of clustering and motions of many cell-surface macromolecules should be achievable.

It is straightforward to combine the two measurements we have described into one scheme in which both the position and the fluorescence power of individual fluorescent objects on

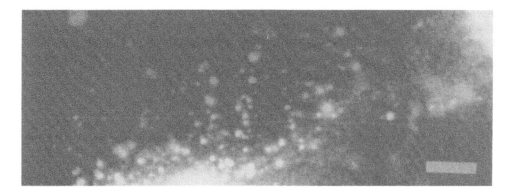

A

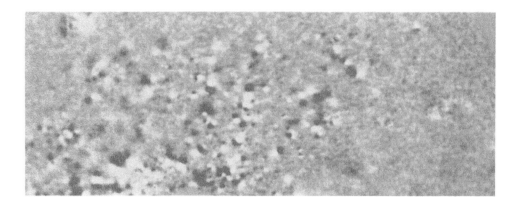

B

FIGURE 12.   The effects of trifluoperazine on LDL-RC mobility. (A) GM2408A treated with 30 μ*M* trifluoperazine. (B) Difference image as in Figure 8A for Δt = 15 sec. Note large motions.

the cell surface can be monitored as a function of time. (However, we have not yet studied the dependence of LDL-RC cluster mobility on cluster size.) The dynamics of various ligand-receptor complexes and clusters visualized on the cell surface can be examined in detail. For example, preliminary data shows coalescence of small clusters of liganded-LDL receptor on GM2408A cells that then remain "stuck" together. We have begun to analyze LDL-RC cluster formation, disintegration, and mobility. Such measurements require second-generation, pattern-recognition software which is currently being developed.

Our results on the clustering of LDL ligand-receptor complexes on the three cell types we have studied are in good agreement with electron microscopy observations and the molecular mobilities are consistent with previous measurements. The observed steady directed motions of liganded-LDL receptors over substantial distances on the cell surface are still not clearly understood, however. Although Bretscher has suggested that flow of lipids from the plasma membrane into coated pits and back out at the leading edges of cells should drag along membrane-bound proteins,[25] we have no evidence that liganded LDL receptors are flowing into focused "vortices" on the cell surface. In fact, we seem to have seen different processes. For example, concerted motion of diI-LDL clusters moving toward the leading edge of a cell (see Figures 7 and 8) is directly opposite to the putative direction of lipid flow. Heath has demonstrated that cross-linked surface proteins on chick embryo fibroblasts are cleared from the lamellipoidal region by cytoskeletal contractile action in a manner that seems qualitatively similar to our observations of LDL motion.[26] We suggest that cytoskeletal activity modulates the mobility of the liganded-LDL receptor and that contractile tractions move receptors. We are examining this hypothesis with the help of extensions of the techniques described in this chapter.

ACKNOWLEDGMENTS

We wish to thank Mary Lipsky and Christine Coulter for their assistance with cell culture and Richik N. Ghosh for his assistance with data collection and with the development of the theory of the uncertainty in single-particle mean-square displacement measurement. This work has been supported by grants from the NSF (PCM 8303404), NIH (GM33028-02), ONR (N00014-84-K-0390), and the Cornell Biotechnology Program.

REFERENCES

1. **Goldstein, J. L., Anderson, R. G. W., and Brown, M. S.,** Coated pits, coated vesicles, and receptor-mediated endocytosis, *Nature (London)*, 279, 679, 1979.
2. **Anderson, R. G. W., Goldstein, J. L., and Brown, M. S.,** Localization of low density lipoprotein receptors on plasma membranes of normal human fibroblasts and their absence in cells from a familial hypercholesterolemia homozygote, *Proc. Natl. Acad. Sci. U.S.A.*, 73, 2434, 1976.
3. **Goldstein, J. L. and Brown, M. S.,** Binding and degradation of low density lipoproteins by cultured human fibroblasts, *J. Biol. Chem.*, 249, 5153, 1974.
4. **Brown, M. S., Dana, S. E., and Goldstein, J. L.,** Regulation of 3-hydroxy-3-methyl-glutaryl coenzyme A reductase activity in cultured human fibroblasts, *J. Biol. Chem.*, 249, 709, 1974.
5. **Brown, M. S. and Goldstein, J. L.,** Analysis of a mutant strain of human fibroblasts with a defect in the internalization of receptor-bound low density lipoprotein, *Cell*, 10, 663, 1976.
6. **Goldstein, J. L. and Brown, M. S.,** Genetics of the LDL receptor: evidence that the mutations affecting binding and internalization are allelic, *Cell*, 12, 692, 1977.
7. **Brown, M. S. and Goldstein, J. L.,** Receptor-mediated endocytosis: insights from the lipoprotein receptor system, *Proc. Natl. Acad. Sci. U.S.A.*, 76, 3330, 1979.
8. **Anderson, R. G. W., Visile, E., Mello, R. J., Brown, M. S., and Goldstein, J. L.,** Immunocytochemical visualization of coated pits and vesicles in human fibroblasts: relation to low density lipoprotein receptor distribution, *Cell*, 15, 919, 1978.
9. **Carpentier, J.-L., Gorden, P., Goldstein, J. L., Anderson, R. G. W., Brown, M. S., and Orci, L.,** Binding and internalization of ^{125}I-LDL in normal and mutant human fibroblasts: a quantitative autoradiographic study, *Exp. Cell Res.*, 121, 135, 1979.
10. **Barak, L. S. and Webb, W. W.,** Fluorescent low density lipoprotein for observation of dynamics of individual receptor complexes on cultured human fibroblasts, *J. Cell Biol.*, 90, 595, 1981.
11. **Barak, L. S. and Webb, W. W.,** Diffusion of low density lipoprotein-receptor complex on human fibroblasts, *J. Cell Biol.*, 95, 846, 1982.
12. **Tank, D. W., Fredericks, W. J., Barak, L. S., and Webb, W. W.,** Electric field-induced redistribution and postfield relaxation of low density lipoprotein receptors on cultured human fibroblasts, *J. Cell Biol.*, 101, 148, 1985.
13. **Sims, P. J., Waggoner, A. S., Wang, C-H., and Hoffman, J. F.,** Studies on the mechanism by which cyanine dyes measure membrane potential in red blood cells and phosphatidylcholine vesicles, *Biochemistry*, 13, 3315, 1974.
14. **Anderson, R. G. W., Brown, M. S., and Goldstein, J. L.,** Inefficient internalization of receptor-bound low density lipoprotein in human carcinoma A-431 cells, *J. Cell Biol.*, 88, 441, 1981.
15. **Schneider, M. B. and Webb, W. W.,** Measurement of submicron laser beam radii, *Appl. Opt.*, 20, 1382, 1981.
16. **Gross, D. and Webb, W. W.,** Molecular counting in small clusters of LDL on cell surfaces by fluorescence intensity quantization, *Biophys. J.*, 45, 269a, 1984.
17. **Gross, D. and Webb, W. W.,** Molecular counting of low density lipoprotein particles in cell surface small clusters, in press *Biophys. J.*, 49, 901, 1986.

18. **Frey, L. D. and Edidin, M.,** The rapid intermixing of cell surface antigens after formation of mouse-human heterokaryons, *J. Cell Sci.*, 7, 319, 1970.

19. **Alexrod, D., Koppel, D. E., Schlessinger, J., Elson, E., and Webb, W. W.,** Mobility measurements by analysis of fluorescence photobleaching recovery kinetics, *Biophys. J.*, 16, 1055, 1976.

20. **Tank, D. W., Wu, E.-S., and Webb, W. W.,** Enhanced molecular diffusibility in muscle membrane blebs: release of lateral constraints, *J. Cell Biol.*, 92, 207, 1982.

21. **Gross, D. and Webb, W. W.,** Time-lapse video recording of individual molecular motions of LDL-receptor complex on living human fibroblasts, *Biophys. J.*, 41, 215a, 1983.

22. **Eisinger, J., Flores, J., and Petersen, W.,** A milling crowd model for local and long-range obstructed lateral diffusion, *Biophys. J.*, 49, 987, 1986.

23. **Gross, D. and Webb, W. W.,** Physical pathways of receptor-mediated endocytosis: post-internalization dynamics of LDL-containing vesicles, in Poster 126, 8th Int. Biophys. Congr., Bristol, U.K., 1984.

24. **Oi, V. T., Glazer, A. N., and Stryer, L.,** Fluorescent phycobiliprotein conjugates for analysis of cells and molecules, *J. Cell Biol.*, 93, 981, 1982.

25. **Bretscher, M. S.,** Endocytosis: relation to capping and cell locomotion, *Science*, 224, 681, 1984.

26. **Heath, J. P.,** Direct evidence for microfilament-mediated capping of surface receptors on crawling fibroblasts, *Nature (London)*, 302, 532, 1983.

27. **Gross, D., Ghosh, R. N., and Webb, W. W.,** unpublished results.

28. **Gross, D. and Webb, W. W.,** unpublished results.

Chapter 11

TOTAL INTERNAL REFLECTION FLUORESCENCE: THEORY AND APPLICATIONS AT BIOSURFACES

Edward H. Hellen, Robert M. Fulbright, and Daniel Axelrod

TABLE OF CONTENTS

I. INTRODUCTION

The distribution and dynamics of molecules at or near surfaces is central to numerous phenomena in biology; e.g., binding to and triggering of cells by hormones, neurotransmitters and antigens; the deposition of plasma proteins upon a foreign surface, leading to thrombogenesis; electron transport; cell adhesion to surfaces; enhancement of the reaction rate with membrane receptors by nonspecific adsorption and surface diffusion of ligand; and the dynamical arrangement of submembrane cytoskeletal structures involved in cell shape, mobility, and mechanoelastic properties.

In most of these examples, certain functionally relevant molecules coexist in both a surface-associated and nonassociated state. If such molecules are detected by a conventional fluorescence technique (such as epi-illumination in a microscope), the fluorescence from the surface-associated molecules may be dwarfed by the fluorescence from nonassociated molecules in the adjacent detection volume. As an optical technique designed to overcome this problem, total internal reflection fluorescence (TIRF) allows selective excitation and detection of just those fluorescent molecules in close (~1000 Å) proximity to the surface.

TIRF is conceptually quite simple. A light beam propagating through a solid (e.g., glass) is directed toward an interface with a liquid (e.g., aqueous solution containing fluorophores) at an angle (measured from the normal) sufficiently large to cause total internal reflection. A very thin (generally less than one light wavelength) "evanescent field" of light thereby develops in the liquid at the interface. This field is capable of exciting fluorophores in the liquid close to the interface while avoiding excitation of a possibly much larger number of fluorophores farther out in the liquid.

This chapter is intended both as a summary of the theoretical and experimental principles of total internal reflection fluorescence, particularly as applied to biophysics, biochemistry, and cell biology, and also as a literature review of very recent work. For reference to previous work, see the review by Axelrod et al.[1] Section II describes some practical optical arrangements for observing TIRF through a microscope; Section III discusses the electromagnetic field which excites TIR fluorescence; Section IV discusses the subsequent emission pattern of a fluorophore near a surface (particularly the interface of water with either a bare dielectric-like glass or a metal-coated dielectric); Section V reviews recent TIRF work on biosurfaces.

II. EXPERIMENTAL ARRANGEMENTS FOR TIRF MICROSCOPY

A wide range of optical arrangements for TIRF have been employed, both with and without a microscope.[1] The arrangements coupled to a microscope are particularly appropriate where small illumination and observation areas are required; e.g., for examining biological cells and for measuring local molecular adsorption kinetic and surface diffusion rates.[2,3] We illustrate four of these microscope adaptations here. All use a visible laser beam source (usually argon or krypton).

Figure 1 shows a possible arrangement for an inverted microscope. This arrangement is easily switched to phase-contrast transmitted illumination or to conventional epi-fluorescence and is also usable with even the shortest working-distance objectives. The key element in the optical system is a fused silica cubical prism that permits the incident laser beam to strike the TIR interface (which may be the surface of a microscope slide or coverslip placed in optical contact with the prism via a drop of glycerol) at greater than the critical angle for total internal reflection. This prism need not be matched in refractive index to the TIR interface material, nor need it be cubical. As an illustration of the optical effect seen with this setup, Figure 2 shows a human erythrocyte ghost labeled with the amphipathic lipid analog 3,3' dioctadecylindocarbocyanine (diI-C_{18}-3), comparing epi-fluorescence with TIRF in the inverted microscope.

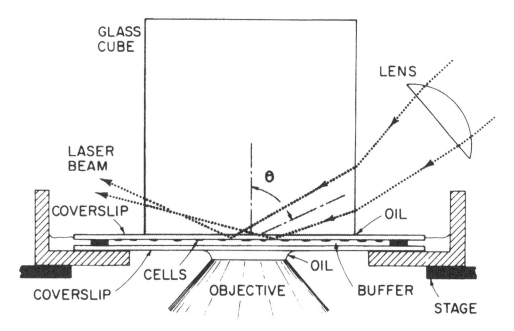

FIGURE 1. TIRF inverted microscope apparatus for viewing cells in culture. Cells are plated and grown on a standard glass coverslip, which is then inverted and placed in optical contact with the cubical prism via a layer of immersion oil or glycerin. (From Axelrod, D., Burghardt, T. P., and Thompson, N. L., *Annu. Rev. Biophys. Bioeng.*, 13, 247, 1984. With permission.)

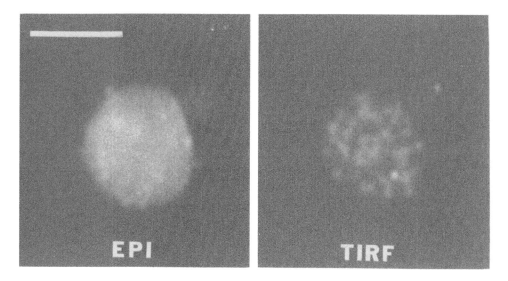

FIGURE 2. Crenated human erythrocyte membrane ghost labeled with diI-C_{18}-(3) and viewed under epi or TIRF illumination. The characteristic spike-like projections of a crenated membrane in close apposition to the glass substrate are much more obvious in the TIRF view. The objective was 100 ×, N.A. = 1.25, oil, immersion. Space bar = 5 μm.

Figure 3 shows a TIRF arrangement for an upright microscope. This setup is particularly appropriate for viewing cultural cells growing in standard plastic culture dishes. The prism is a trapezoid (actually, a truncated 60° equilateral triangle made of high refractive index flint glass) brought into optical contact (via a drop of glycerol) with the bottom of the culture dish. This arrangement has the advantage that: (1) the culture dish can be inserted and

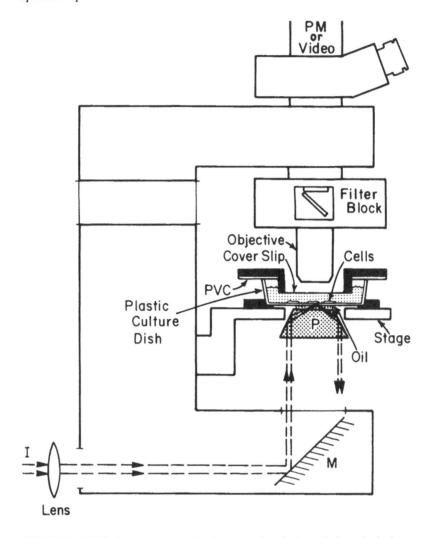

FIGURE 3. TIRF adapted to an upright microscope, for viewing cells in a plastic tissue culture dish. The prism is a truncated equilateral triangle. The region of the sample chamber is shown enlarged relative to the rest of the microscope for pictorial clarity. Abbrevations are as follows: I, incident light; M, mirror (part of microscope base); P, prism; PM, photomultiplier; PVC, polyvinylidene chloride ("Saran"® film used to seal in a 10% CO_2 atmosphere over the tissue culture medium). (From Axelrod, D., Burghardt, T. P., and Thompson, N. L., *Annu. Rev. Biophys. Bioeng.*, 13, 247, 1984. With permission.)

removed easily; and (2) the illuminated region does not move when the microscope is focused. It has the disadvantage that the incidence angle is not variable.

Figure 4 shows a TIRF system for either an upright or inverted microscope. It allows the incidence angle to be changed continuously and allows two TIR beams to intersect at the interface at any relative angle, thereby creating an interference fringe pattern of illumination of variable interfringe spacing. Figure 5 shows such a fringe pattern generated by this setup on diI-C_{18}-(3)-labeled erythrocyte ghost membrane flattened onto a glass surface. This fringe pattern is useful for studies of surface diffusion, as discussed in Section V.D. The interfringe spacing can easily be shown to equal $\lambda_0 (2n \sin \theta \sin \alpha/2)^{-1}$ where λ_0 is the light wavelength in vacuum, n is the refractive index of the solid, θ is the incidence angle measured from the normal, and α is the intersection angle of the two beams in the plane of the TIRF surface. Given the known or easily measured parameters λ_0, n, and α, one can measure the interfringe

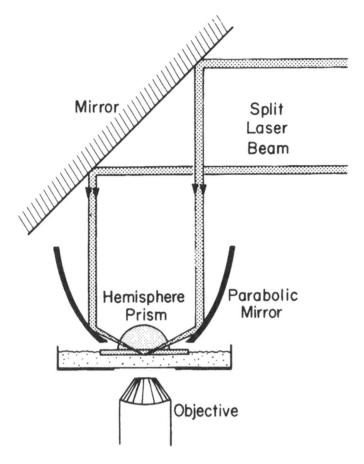

FIGURE 4. Intersecting TIR beams split from the same laser to produce interference fringes for viewing in an inverted microscope. The two beams reflect at a paraboloidal mirror, enter a hemispherical prism, and reconverge at the TIRF surface, which is positioned at the focus of the parabola. The beams must be coherent; i.e., the total path difference must be less than the coherence length of the laser. The beams need not reflect at opposite sides of the paraboloid, but rather at any relative azimuthal angle around the paraboloid, thereby allowing adjustment of the intersection angle.

spacing and thereby calculate θ, which is often difficult to directly measure exactly. TIRF interference fringes were first introduced by Weis et al.[4]

A fourth illumination system is depicted in Figure 6. It is perhaps the least flexible of all, but it requires no prism whatsoever. A high aperture objective epi-illuminates the sample with a beam that propagates near the edge of the objective's pupil and thereby strikes the interface at a highly oblique angle. This system requires an objective with a numerical aperture of at least 1.33 (the refractive index of water) for the most extreme oblique rays to be incident at greater than the critical angle for a glass/water interface.

In setting up a TIRF system, one may encounter a number of questions about design and materials. The following suggestions may be helpful:

1. The prism used to couple the light into the system and the (usually disposable) slide or coverslip in which the total reflection takes place need not be exactly matched in refractive index.
2. The prism and slide may be optically coupled with glycerin, cyclohexanol, or micro-

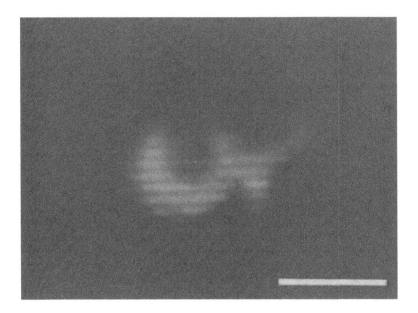

FIGURE 5.   Interference pattern formed by two TIRF beams intersecting each other at a 20° angle in the optical arrangement depicted in Figure 4. The sample is a flattened erythrocyte ghost prepared in the manner described in Section V.D and labeled with rhodamine wheat germ agglutinin. The laser wavelength is 514.5 nm (in vacuum), and the resulting fringe spacing is approximately 1.1 μm. The objective is 50 ×, N.A. = 100, water, immersion. Space bar = 10 μm.

scope-immersion oil, among other liquids. Immersion oil has a higher index of refraction (thereby avoiding possible total internal reflection at the prism/glycerin interface for low incidence angles), but tends to be more autofluorescent (even the "extremely low" fluorescence types). This problem is usually not important to TIRF microscopy, but can be serious in large-area applications.

3.   The prism and slide can both be made of ordinary optical glass for many applications, unless shorter penetration depths arising from higher refractive indexes are desired. (More exotic high-refractive index materials such as sapphire, titanium dioxide, and strontium titanate can yield penetration depths as low as $\lambda_0/20$.) However, optical glass does not transmit light below about 310 nm and also has a dim luminescence with a long (several hundred microsecond) decay time, which can be a problem in some photobleaching experiments. The luminescence of high-quality fused silica (often called "quartz") is much lower.

4.   The total internal reflection surface need not be polished to a higher degree than a standard commercial microscope slide.

5.   Either a laser or conventional arc light source will suffice for study of macroscopic areas. But in TIRF microscopy, a conventional light source is difficult to focus to a small enough region at high enough intensity, while still retaining sufficient collimation to avoid partial light transmission through the interface; a laser is more desirable here.

6.   Illumination of surface-adsorbed proteins can lead to apparent photochemically induced cross-linking and also to photobleaching at the higher range of intensities that might be used in TIRF studies. Apparent cross-linking, measured as a slow, continual, illumination-dependent increase in observed fluorescence, can be inhibited by 0.05 *M* cysteamine, among other substances; photobleaching can be reduced somewhat by deoxygenation or by 0.01 *M* sodium dithionite, among other substances.

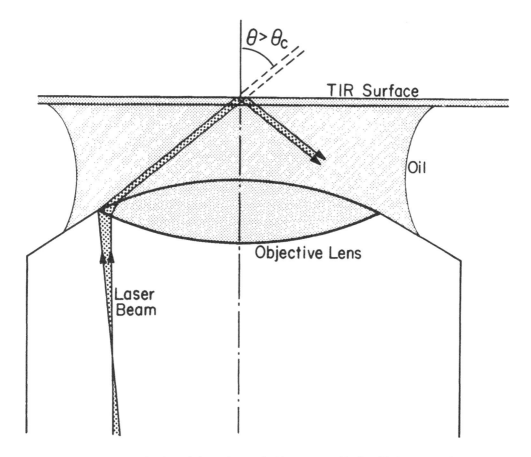

FIGURE 6.   TIR illumination through the periphery of a high-aperture objective. We have successfully tried this method with a Zeiss® 63 ×, N.A. = 1.4, oil immersion objective, but the sample region illuminated by TIR covers only a fairly small area near the center of the field of view. The vertical scale here is somewhat stretched for clarity. Incidence angle θ will be greater than critical angle θ_c only for extreme peripheral rays.

III. EXCITATION OF FLUOROPHORES BY TIR

A. Single Interface

When a light beam propagating through a transparent medium 3 of high index of refraction (e.g., a solid-glass prism) encounters an interface with a medium 1 of a lower index of refraction (e.g., an aqueous solution), it undergoes total internal reflection for incidence angles (measured from the normal to the interface) greater than the "critical angle". The critical angle θ_c, is given by:

$$\theta_c = \sin^{-1}(n_1/n_3)$$

$$= \sin^{-1}(\epsilon_1/\epsilon_3)^{1/2} \tag{1}$$

where n_1 and n_3 are the refractive indexes of the liquid and the solid, respectively, and ϵ_1 and ϵ_3 are the corresponding dielectric constants. Although the incident light beam totally internally reflects at the interface, an electromagnetic field called the "evanescent wave" penetrates a small distance into the liquid medium and propagates parallel to the surface in the plane of incidence. The evanescent wave is capable of exciting fluorescent molecules that might be present near the interface.

The evanescent electric-field intensity I(z) decays exponentially with perpendicular distance z from the interface:

$$I(z) = I(0) \, e^{-z/d} \tag{2}$$

where:

$$d = \frac{\lambda_0}{4\pi} [\epsilon_3 \sin^2\theta - \epsilon_1]^{-1/2} \tag{3}$$

for angles of incidence $\theta > \theta_c$ and light wavelength in vacuum λ_0. Depth d is independent of the polarization of the incident light and decreases with increasing θ. Except for $\theta \simeq \theta_c$ (where $d \rightarrow \infty$), d is on the order of λ_0 or smaller.

It is important to examine the electric and magnetic fields of the TIR evanescent wave because of their remarkable polarization and phase behavior and because the expressions are needed for the calculations of I(0) in Equation 2. The field components are listed below, with incident electric-field amplitudes $A_{p,s}$ and phase factors relative to those of the incident **E*** and **H** fields' phase at $z = 0$. (The coordinate system is chosen such that the x-z plane is the plane of incidence.)

$$E_x = \left[\frac{2 \cos\theta (\sin^2\theta - n^2)^{1/2}}{(n^4 \cos^2\theta + \sin^2\theta - n^2)^{1/2}} \right] A_p e^{-i(\delta_p + \pi/2)} \tag{4}$$

$$E_y = \left[\frac{2 \cos\theta}{(1 - n^2)^{1/2}} \right] A_s e^{-i\delta_s} \tag{5}$$

$$E_z = \left[\frac{2 \cos\theta \sin\theta}{(n^4 \cos^2\theta + \sin^2\theta - n^2)^{1/2}} \right] A_p e^{-i\delta_p} \tag{6}$$

where:

$$\delta_p \equiv \tan^{-1} \left[\frac{(\sin^2\theta - n^2)^{1/2}}{n^2 \cos\theta} \right] \tag{7}$$

$$\delta_s \equiv \tan^{-1} \left[\frac{(\sin^2\theta - n^2)^{1/2}}{\cos\theta} \right] \tag{8}$$

and:

$$n \equiv n_1/n_3 = (\epsilon_1/\epsilon_3)^{1/2}$$

Note that the evanescent electric field is transverse to the propagation direction ($+x$) only for the s polarization. The p- polarized **E** field "cartwheels" along the surface with a spatial period of $\lambda_0/(n_3 \sin \theta)$ as shown schematically in Figure 7.

For absorbers with magnetic dipole transitions, the evanescent magnetic field **H** is relevant. Assuming equal magnetic permeabilities at both sides of the interface, the components of the evanescent field **H** at $z = 0$ are

$$H_x = \left[\frac{2 \cos\theta (\sin^2\theta - n^2)^{1/2}}{(1 - n^2)^{1/2}} \right] A_s e^{-i(\delta_s - \pi)} \tag{9}$$

$$H_y = \left[\frac{2n^2 \cos\theta}{(n^4 \cos^2\theta + \sin^2\theta - n^2)^{1/2}} \right] A_p e^{-i(\delta_p - \pi/2)} \tag{10}$$

---

*   Vectors are indicated by boldface type.

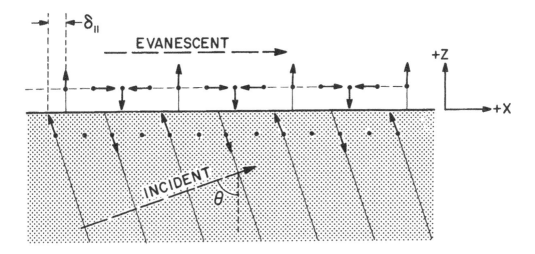

FIGURE 7. Electric field vectors of incident and evanescent light for the ‖ incident polarization, showing the phase lag $\delta_{\|}$ and the "cartwheel" or elliptical polarization of the evanescent field in the plane of propagation. Both the incident and evanescent vectors refer to the z = 0 position; they are schematically displaced a short distance below and above the interface along their constant phase lines for pictorial clarity only.

$$H_z = \left[\frac{2\cos\theta\,\sin\theta}{(1 - n^2)^{1/2}}\right] A_s e^{-i\delta_s} \tag{11}$$

The angular dependence of the phase factors δ_s and δ_p gives rise to a measurable longitudinal shift of a finite-size incident beam, known as the Goos-Hanchen shift. Viewed physically, some of the energy of a finite-width beam crosses the interface into the lower refractive index material, skims along the surface for a Goos-Hanchen shift distance ranging from a fraction of a wavelength (at $\theta = 90°$) to infinite (at $\theta = \theta_c$), and then reenters the higher refractive index material.

In many TIRF experiments, an incident laser beam of Gaussian intensity profile passes through a converging lens, enters the side of a prism, and then focuses at a totally reflecting surface. The nature of the evanescent illumination produced by such a focused-finite beam geometry can be calculated by mathematically expressing the incident convergent beam as an integral over planes waves propagating at different angles. For the range of convergence angles usually encountered in TIRF, the evanescent illumination is of an elliptical Gaussian profile and the polarization and penetration depth are approximately equal to those of a single plane wave.[5]

The average energy flux in the evanescent wave is given by the real part of the Poynting vector $\mathbf{S} = \left(\dfrac{c}{4\pi}\right) \mathbf{E} \times \mathbf{H}^*$. However, the probability of absorption of energy per unit time from the evanescent wave by an electric dipole-allowed transition of moment $\boldsymbol{\mu}$ in a fluorophore is proportional to $|\boldsymbol{\mu} \cdot \mathbf{E}|^2$. Note that Re $\mathbf{S}$ and $|\boldsymbol{\mu} \cdot \mathbf{E}|^2$ for an evanescent wave have a different dependence on θ.

Given randomly oriented dipoles, the absorption probability rate is proportional to the "intensity" $I_{p,s} \xi |E_{p,s}|^2$ for evanescent polarizations parallel (p) or perpendicular (s) to the plane of incidence, where $\mathbf{E}_p = E_x\,\hat{x} + E_z\,\hat{z}$ and $\mathbf{E}_s = E_y\,\hat{y}$. At z = 0 these intensities are

$$I_p(0) = \mathcal{I}_p \cdot \frac{4\cos^2\theta(2\sin^2\theta - n^2)}{n^4\cos^2\theta + \sin^2\theta - n^2} \tag{12}$$

and:

$$I_s(0) = \mathscr{I}_s \cdot \frac{4 \cos^2\theta}{1 - n^2} \tag{13}$$

and $\mathscr{I}_{p,s} \xi |A_{p,s}|^2$ is the intensity of the incident electric field in the glass.

Figure 8a illustrates $I_{p,s}(0)$ as functions of θ. Note several features: (a) the evanescent intensity $I_{p,s}(0)$ is not weak and can be several times stronger than the incident intensity for angles of incidence within a few degrees of the critical angle; (b) the intensities of the two polarizations are different, with $I_p(0)$ somewhat more intense for all θ; (c) $I_{p,s}(0)$ both approach zero as $\theta \rightarrow 90°$.

B. Intermediate Layer

In a variety of experimental circumstances, the TIR surface may not be a simple interface between two media, but rather a stratified multilayer system. Biophysically relevant examples of a third medium sandwiched between the glass and water might include a planar phospholipid monolayer or bilayer (as a model-membrane surface) or a vacuum-deposited coating of semiconductor or metal (employed to quench fluorescence very near the surface, as discussed in Section IV). We briefly discuss the TIR evanescent wave in a three-layer system here.

The incident beam propagates through medium 3 (usually glass) and encounters medium 2 of arbitrary thickness δ (possibly lipid or metal) at incidence angle θ. If $\theta > \theta_c$ (as defined in Equation 1), the field **E** created in medium 1 will be evanescent. Using the same coordinate system as previously, with the origin at the medium 2/medium 1 interface and the $+z$ axis directed into medium 1, we can write the $z = 0$ field components with their phases relative to that of the incident electric field $A_{p,s}$ at $z = 0$:

$$E_x = -A_p \, \alpha_{13} \, T_p \tag{14}$$

$$E_y = A_s T_s \tag{15}$$

$$E_z = A_p(\epsilon_3/\epsilon_1)^{1/2} \sin\theta \, T_p \tag{16}$$

$T_{p,s}$ are the Fresnel coefficients for transmission through a stratified three-medium system with the beam incident from the medium 3 side:[6]

$$T_{p,s} = \frac{t_{32}^{p,s} \, t_{21}^{p,s} \, e^{ik_1\delta\alpha_{23}}}{1 + r_{32}^{p,s} \, r_{21}^{p,s} \, e^{2ik_1\delta\alpha_{23}}} \tag{17}$$

where t_{32}, t_{21}, r_{32}, and r_{21} are the transmission and reflection Fresnel coefficients for a single interface, listed for convenience here:

$$r_{ij}^p = \frac{\epsilon_j\alpha_{i1} - \epsilon_i\alpha_{j1}}{\epsilon_j\alpha_{i1} + \epsilon_i\alpha_{j1}} \tag{18}$$

$$t_{ij}^p = \frac{2(\epsilon_i\epsilon_j)^{1/2}\alpha_{i3}}{\epsilon_j\alpha_{i3} + \epsilon_i\alpha_{j3}} \tag{19}$$

$$r_{ij}^s = \frac{\alpha_{i1} - \alpha_{j1}}{\alpha_{i1} + \alpha_{j1}} \tag{20}$$

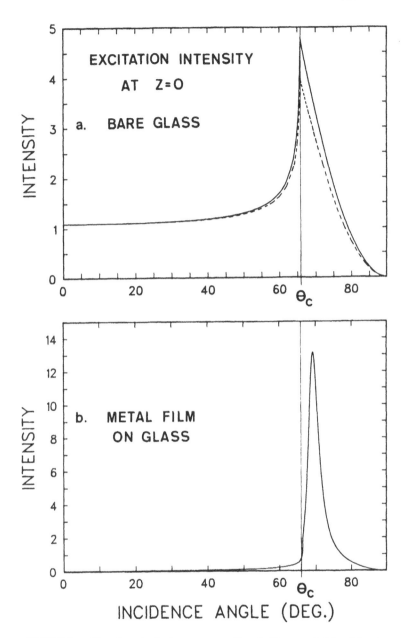

FIGURE 8a and b.   Intensity $I_{p,s}$ in water at z = 0 vs. incidence angle θ for dielectric constants ϵ_1 = 1.77, ϵ_3 = 2.13 (indexes of refraction n_1 = 1.33, n_3 = 1.46) corresponding to a critical angle of θ_c = 65.7°. The incident intensity in medium 3 (the glass) is $\mathcal{I}_{p,s}$ = 1. (a) I_p (solid line) and I_s (dashed line) for bare glass; (b) I_p for a 20 nm aluminum intermediate layer. Note that I_s is too small to be apparent on this graph. The incident light has vacuum wavelength λ_0 = 520 nm, for which the aluminum is characterized by ϵ_2 = −32.5 + i 8.4 ($\hat{n}_2$ = 0.73 + i 5.75). Note the difference in ordinate scale between (a) and (b).

$$t_{ij}^s = \frac{2\alpha_{i3}}{\alpha_{i3} + \alpha_{j3}} \tag{21}$$

The α_{ij} factors are

$$\alpha_{ij} \equiv \left(\frac{\epsilon_i}{\epsilon_1} - \frac{\epsilon_j}{\epsilon_1} \sin^2\theta \right)^{1/2} \tag{22}$$

Also, $k_1 \equiv \frac{\omega}{c}\epsilon_1^{1/2}$ where ω is the excitation light angular frequency.

Note that α_{ij} and $T_{p,s}$ can be complex, so the amplitudes and phase factors analogous to those of Equations 4 to 6 are only implicit here. Nevertheless, Equations 14 to 16 are a generalization of the single-interface Equations 4 to 6 and reduce to them as expected if we set $\epsilon_2 = \epsilon_3$.

The intensity of the evanescent wave is

$$I_p = A_p^2 |T_p|^2 \left(2 \frac{\epsilon_3}{\epsilon_1} \sin^2\theta - 1 \right) e^{-z/d} \tag{23}$$

$$I_s = A_s^2 |T_s|^2 e^{-z/d}$$

The qualitative behavior of I given by Equation 23 is noteworthy. First, the exponential decay of the evanescent intensity has exactly the same characteristic distance d as does the single-interface case (Equation 2), despite the intervening layer of arbitrary material and thickness. Second, the dependence of I_p upon θ (Figure 8b) exhibits a rather striking peak if the intermediate layer is a metal. A metal has a dielectric constant consisting of a negative real part and a positive imaginary part. (For aluminum, $\epsilon_2 = -32.5 + 8.4 i$ at $\lambda_0 = 520$ nm). At a certain angle of incidence θ_p (such that $\frac{\pi}{2} > \theta_p > \theta_c$), the denominator of t_{21}^p becomes quite small (due to the oppositely signed real parts of ϵ_2 and ϵ_1). At this angle, E_x and E_z can become very large, in some cases much larger than the incident-field amplitude. This resonance-like effect is due to excitation of a surface plasmon mode at the metal/water interface.[7-9]

IV. EMISSION BY FLUOROPHORES NEAR A SURFACE

A. Introduction

Although the probability of absorption of TIR evanescent energy by a fluorophore of given orientation decreases exponentially with distance z from a dielectric surface, the intensity of fluorescence actually viewed by a detector varies with z in a much more complicated fashion. Both the angular pattern of emitted radiation and the fluorescent lifetime are altered as a function of z by the proximity of the surface. Various aspects of fluorophore emission near surfaces have been discussed.[8-15,33] We present here a condensed theoretical explanation and quantitative summary of original calculations particularly appropriate for use in TIRF.

In many TIRF applications, the exact z dependence of the observed fluorescence intensity is irrelevant, as long as it decays rapidly enough to selectively report only surface-bound fluorophores. But in other cases, knowledge of the z-dependence is needed; e.g., to determine the fluorophore concentration as a function of z; to determine the distance between the TIR surface and a layer of fluorophore (such as a labeled membrane); or to predict the behavior of special surfaces (such as metal coatings that quench fluorescence).

In general, the fluorophores have a two-dimensionally homogeneous concentration $C(\Omega'$, z) where Ω' represents the polar and azimuthal angles θ' and ϕ' of the emission transition electric dipole moment μ', with the coordinate axes in the same directions as used in Section III. We will assume that the illumination is an evanescent wave formed by an infinite plane wave incident at angle θ, and that the fluorophore concentration is low enough to leave the evanescent wave only negligibly perturbed.[3]

We adopt the same three-layer surface model as in Section III.B. A fluorophore resides in medium 1 (water) at a positive distance z from an interface with an intermediate layer of medium 2 (metal, liquid, or glass) which extends from $z = 0$ down to $z = -\delta$. Substrate medium 3 (glass) extends from $z = -\delta$ to $z \rightarrow -\infty$. Where necessary, numerical subscripts on a parameter refer to the indicated medium.

Rather than calculate the fluorescence intensity at each observation point $\mathbf{r}$, we write expressions for fluorescence power $P_c(\beta_0)$ integrated over all azimuthal observation angles ϕ $(0 \rightarrow 2\pi)$ and a specified range of polar angles β. The polar angle β is measured from the $+z$ axis when integrating the power emitted into the water. β is measured from the $-z$ axis when integrating the power emitted into the glass. In each case the range of β is $0 \rightarrow \beta_0$. $P_c(\beta_0)$ is proportional to the fluorescence power observed by a microscope objective of aperture angle β_0*:

$$P_{c,i}^{p,s}(\beta_0) = K \, \epsilon_i^{1/2} \int |\hat{E}'|^2 \, |\mu \cdot E_{p,s}|^2 \, C(\Omega', z) \, dz \, d\Omega' \, \sin\beta \, d\beta \, d\phi \qquad (24a)$$

where

K = a product of unit conversion factors, fundamental and numerical constants, quantum and light collection efficiencies, etc.**

ϵ_i = the dielectric constant of medium i in which the observation point is embedded. The dielectric constant appears here due to the Poynting vector expression for the intensity of propagating light. The magnetic permittivity is assumed equal to unity and is not shown.

$\hat{E}'$ = $\hat{E}'(\Omega', z)$ is the electric field at observation point (R, β, ϕ) produced by a dipole μ' of orientation $\Omega' = (\theta', \phi')$ at distance z. $\epsilon_i^{1/2} |\hat{E}'|^2$ is normalized to represent the local light intensity *per* unit of total power P_t dissipated by the dipole. The superscript ^ refers to the normalization.

μ = the absorption-transition dipole, assumed parallel to the emission dipole μ' here. Parallelness is assumed for simplicity, but it is a good approximation for tetramethylrhodamine, fluorescein, and carbocyanine dyes excited in the visible range.

$E_{p,s}$ = $E_{p,s}(\theta, z)$ is the electric field of the TIR evanescent wave for p- or s- polarized incident light. The z- exponentially decaying explicit form of $E_{p,s}$ is given in Section III.

$d\Omega'$ = $\sin \theta' \, d\theta' \, d\phi'$ represents the orientation angle differential for the emission dipole μ'.

It is useful to write P_c in other forms, which differ only in the order in which the multiple integration is performed:

---

* The numerical aperture of the objective is $n_i \sin \beta_0$ where n_i is the refractive index of medium i. (i = 1 or 3 for observation from the water or glass side, respectively). We neglect any angle-dependent reflections in the microscope objective system.

** The quantum efficiency of emission may depend on z due to perturbation of the molecular orbital quantum states of the fluorophore by the local electrostatic environment of the surface, but we neglect such effects here.

$$P_{c,i}^{p,s}(\beta_0) = K \int_0^{\beta_0} F_i^{p,s}(\beta) \sin\beta \, d\beta \tag{24b}$$

$$= K \int \hat{Q}_i(\beta_0, \theta', z) \, |\boldsymbol{\mu} \cdot \mathbf{E}^{p,s}|^2 \, C(\Omega', z) \, dz \, d\Omega' \tag{24c}$$

$$= K \int \hat{S}_i(\beta, \theta', z) \, |\boldsymbol{\mu} \cdot \mathbf{E}^{p,s}|^2 \, C(\Omega', z) \, dz \, d\Omega' \sin\beta \, d\beta \tag{24d}$$

The mathematical definitions of F, $\hat{Q}$, and $\hat{S}$ are clear from a comparison of Equations 24b, c, or d with Equation 24a. Physically, F is the ϕ-integrated intensity observed at polar angle β emitted from a TIR-excited layer of fluorophore. $\hat{Q}$ is the ratio of power emitted by a dipole of polar orientation angle θ' and distance z into a given microscope objective of aperture angle β_0, to the total power dissipated by the dipole. $\hat{S}_i$ is the ϕ-integrated emission intensity per unit of total dissipated power emitted by a dipole oriented at polar angle θ' and distance z and observed at polar angle β.

Even in the case where the concentration $C(\Omega', z)$ depends only on z and not on Ω', P_c is not a simple Laplace transform of $C(z)$, because of the presence of ratio $\hat{Q}_i$ (see Equation 24c). $\hat{Q}_i$ is a rather complicated function of z, neither a constant nor an exponential for any range of angles of observation. Nevertheless, information about $C(z)$ can still be obtained from Equation 24 because $|\hat{E}'|^2$ and ultimately $\hat{Q}_i$ can be derived from electromagnetic theory, as outlined below.

We start by modeling the quantum mechanical emission dipole by a continuously oscillating dipole $\boldsymbol{\mu}'$. The magnitude of $\boldsymbol{\mu}'$ is not necessarily fixed for all orientations and distances z from the surface, but must be determined according to the rate of energy absorption and dissipation, assumed to be in steady state.

An *isolated* dipole $\boldsymbol{\mu}'$ oscillating with frequency ω' in a homogeneous medium 1 produces a field $\boldsymbol{\epsilon}$ at observation point $\mathbf{r}$ which can be written as a Fourier integral of plane waves over all $\mathbf{k}$.[8]

$$\boldsymbol{\epsilon}(\mathbf{r}) = -\frac{e^{-i\omega't}}{2\pi^2\epsilon_1} \int d\mathbf{k} \, e^{i\mathbf{k}\cdot(\mathbf{r}-\mathbf{r}_0)} \left[\boldsymbol{\mu}' + \frac{\mathbf{k} \times (\mathbf{k} \times \boldsymbol{\mu}')}{k^2 - k_1^2}\right] \tag{25}$$

where:

$\mathbf{r}_0 = z\hat{z}$ is the position of dipole $\boldsymbol{\mu}'$ $k_1 = \dfrac{\omega'}{c} \epsilon_1^{1/2}$ is the magnitude of the wave vector of propagating waves in medium 1.

Equation 25 has interesting features which have a physical interpretation. First, the integration ranges over only those $\mathbf{k}$ such that $\text{Re}[\mathbf{k} \cdot (\mathbf{r} - \mathbf{r}_0)]$ is positive, corresponding to waves propagating away from the dipole. Second, the denominator in the integrand effectively ensures that the only $\mathbf{k}$ vectors significantly contributing to the integral will have $|\mathbf{k}| = k_1$. This mathematical feature corresponds to the physical requirement that all of the plane waves have the same frequency ω'. Nevertheless, the integration allows two of the components of $\mathbf{k}$, say k_x and k_y, to range over *all* real values, provided that k_z is allowed to become imaginary for $(k_x^2 + k_y^2) > k_1^2$. Wavevectors $\mathbf{k}$ with imaginary k_z correspond to fields propagating away from the dipole in the x-y plane at distance z but exponentially decaying in both z-directions from that plane. These decaying fields are collectively called the "near field" of the dipole.

Now we place the isolated dipole of Equation 25 near a surface. Each of the plane waves represented in the integrand of Equation 25 that propagate toward the surface, as well as the exponentially decaying fields extending in the negative z direction, interact with the surface and suffer an alteration of magnitude and phase during reflection or transmission, as given by Fresnel coefficients. The actual field produced at any distant **r** in medium 1 can then be expressed as a superposition of direct and reflected fields integrated over **k** with the relative weightings given by Equation 25. Likewise, the distant field in medium 3 is an integral representing the superposition of transmitted fields.

Physically, an exponentially decaying field of the dipole interacts significantly with the surface only if the dipole is within the characteristic field decay distance $|k_z|^{-1}$ of the surface. Some of the decaying fields satisfy $k_1^2 < (k_x^2 + k_y^2) < k_3^2$ (where $k_3 = (\omega'/c)\,\epsilon_3^{1/2}$). These fields generate propagating waves in medium 3 which travel away from the surface at greater than the critical angle. It is thereby clear why close proximity to a surface both increases the overall rate of emission of a dipole and also biases it toward increased energy flow into the higher refractive index medium.

In the next subsection, we outline the steps in the calculation. However, the reader may skip directly to Subsection C (Conclusions).

B. Calculations

We are interested in the electric field at observation point (R, β, ϕ) in spherical coordinates. The integration is somewhat simpler if we define a new coordinate system (x', y', z') rotated about the z-axes by an angle of ϕ, such that, the observation point is in the $y' = 0$ plane at polar angle β. The details of the integration are given by Hellen and Axelrod.[32]

Table 1 presents the results of calculations for electric fields **E′** from dipoles of magnitude μ' oriented along each of the three axes $(x', y',$ and $z')$, and observed at a large distance R in either medium 1 (water) or medium 3 (glass). The results involve the Fresnel reflection and transmission coefficients $r_{p,s}$ and $t_{p,s}$ for a stratified three-medium system with the beam incident from the medium 1 side. These are

$$r_{p,s} = \frac{r_{12}^{p,s} + r_{23}^{p,s}\,e^{2ik_1\delta\alpha_{21}}}{1 + r_{12}^{p,s}\,r_{23}^{p,s}\,e^{2ik_1\delta\alpha_{21}}} \tag{26}$$

$$t_{p,s} = \frac{t_{12}^{p,s}\,t_{23}^{p,s}\,e^{2ik_1\delta\alpha_{23}}}{1 + r_{12}^{p,s}\,r_{23}^{p,s}\,e^{2ik_1\delta\alpha_{23}}} \tag{27}$$

where r_{12}, r_{23}, t_{12}, and t_{23} have already been defined in Equations 18 to 21, except in this case with:

$$\alpha_{ij} \equiv \left(\frac{\epsilon_i}{\epsilon_1} - \frac{\epsilon_j}{\epsilon_1}\sin^2\beta\right)^{1/2} \tag{28}$$

The radiated power per unit area normal to the observation direction is given by the real part of the Poynting vector, $\dfrac{c}{8\pi}\left(\dfrac{\epsilon_i}{\mu_i}\right)^{1/2}|\mathbf{E}'|^2$, where $i = 1$ (water) or 3 (glass) and the magnetic permittivities μ_i are assumed equal to unity. Starting from the expressions for **E′** in Table 1, we can easily calculate $|\mathbf{E}'|^2$. The results are displayed in Table 2.

To bring these results for $|\mathbf{E}'|^2$ into a form useable in Equation 24a, two modifications are necessary: (1) generalization to an arbitrarily oriented dipole and integration over azimuthal observation angle ϕ; and (2) normalization with respect to total dissipated power (which, for a given dipole magnitude $|\mu'|$, is a function of both dipole orientation polar angle θ' and distance z).

Table 1
CALCULATED ELECTRIC FIELD COMPONENTS

Dipole orientation	E' Component	Observed in	
		Water	Glass
$\mu' = \mu'\hat{x}$	E'_x	$\dfrac{\mu' k_1^2}{\epsilon_1} \dfrac{e^{ik_1R}}{R} (\cos^2\beta)\,(e^{-ik_1z\cos\beta} - r_p e^{ik_1z\cos\beta})$	$\dfrac{\mu' k_3^2}{(\epsilon_1\epsilon_3)^{1/2}} \dfrac{e^{ik_3R}}{R} (\cos^2\beta)\, t_p e^{ik_3\beta\cos\beta} e^{ik_1z\alpha_{13}}$
	E'_y	0	0
	E'_z	$\dfrac{-\mu' k_1^2}{\epsilon_1} \dfrac{e^{ik_1R}}{R} (\cos\beta\sin\beta)\,(e^{-ik_1z\cos\beta} - r_p e^{ik_1z\cos\beta})$	$\dfrac{\mu' k_3^2}{(\epsilon_1\epsilon_3)^{1/2}} \dfrac{e^{ik_3R}}{R} (\cos\beta\sin\beta)\, t_p e^{ik_3\beta\cos\beta} e^{ik_1z\alpha_{13}}$
$\mu' = \mu'\hat{y}$	E'_x	0	0
	E'_y	$\dfrac{\mu' k_1^2}{\epsilon_1} \dfrac{e^{ik_1R}}{R} (e^{-ik_1z\cos\beta} + r_s e^{ik_1z\cos\beta})$	$\dfrac{\mu' k_3^2}{(\epsilon_1\epsilon_3)^{1/2}} \dfrac{e^{ik_3R}}{R} \dfrac{\cos\beta}{\alpha_{13}}\, t_s e^{ik_3\beta\cos\beta} e^{ik_1z\alpha_{13}}$
	E'_z	0	0
$\mu' = \mu'\hat{z}$	E'_x	$\dfrac{-\mu' k_1^2}{\epsilon_1} \dfrac{e^{ik_1R}}{R} (\cos\beta\sin\beta)\,(e^{-ik_1z\cos\beta} + r_p e^{ik_1z\cos\beta})$	$\dfrac{\mu' k_3^2}{\epsilon_1} \dfrac{e^{ik_3R}}{R} \dfrac{\cos^2\beta\sin\beta}{\alpha_{13}}\, t_p e^{ik_3\beta\cos\beta} e^{ik_1z\alpha_{13}}$
	E'_y	0	0
	E'_z	$\dfrac{\mu' k_1^2}{\epsilon_1} \dfrac{e^{ik_1R}}{R} \sin^2\beta\,(e^{-ik_1z\cos\beta} + r_p e^{ik_1z\cos\beta})$	$\dfrac{\mu' k_3^2}{\epsilon_1} \dfrac{e^{ik_3R}}{R} \dfrac{\cos\beta\sin^2\beta}{\alpha_{13}}\, t_p e^{ik_3\beta\cos\beta} e^{ik_1z\alpha_{13}}$

Table 2
CALCULATIONS OF $|E'|^2$

Dipole orientation	$|E'|^2$ Observed in	
	Water	**Glass**
$\mu' = \mu'\hat{x}$	$\dfrac{\mu'^2 k_1^4}{\epsilon_1^2 R^2} \cos^2\beta \, |e^{-2ik_1 z\cos\beta} - r_p|^2$	$\dfrac{\mu'^2 k_3^4}{\epsilon_1\epsilon_3 R^2} \cos^2\beta \, |t_p e^{ik_1 z\alpha_{13}}|^2$
$\mu' = \mu'\hat{y}$	$\dfrac{\mu'^2 k_1^4}{\epsilon_1^2 R^2} |e^{-2ik_1 z\cos\beta} + r_s|^2$	$\dfrac{\mu'^2 k_3^4}{\epsilon_1\epsilon_3 R^2} \cos^2\beta \, \left|\dfrac{t_s}{\alpha_{13}} \, e^{ik_1 z\alpha_{13}}\right|^2$
$\mu' = \mu'\hat{z}$	$\dfrac{\mu'^2 k_1^4}{\epsilon_1^2 R^2} \sin^2\beta \, |e^{-2ik_1 z\cos\beta} + r_p|^2$	$\dfrac{\mu'^2 k_3^4}{\epsilon_1^2 R^2} \cos^2\beta \sin^2\beta \, \left|\dfrac{t_p}{\alpha_{13}} \, e^{ik_1 z\alpha_{13}}\right|^2$

1. Arbitrary Dipole Orientation

An arbitrarily oriented dipole μ' can be decomposed by projection into three dipoles oscillating, in phase, along each of the three primed axes. The electric field $E'(\mu')$ produced by μ' is simply the superposition of electric fields (as listed in Table 1) from each of the three dipole components with their relative magnitudes given by the projected components of the original dipole. After forming this more general $|E'|^2$ and then integrating over observation azimuthal angle ϕ, we obtain the ϕ-integrated emission intensity S_i propagating at observation polar angle β from a dipole oriented at polar angle θ', in terms of the expressions for $|E'|^2$ for dipoles oriented along $\hat{x}'$, $\hat{y}'$, and $\hat{z}'$ (as in Table 2):

$$S_i(\theta', \beta, z) \equiv \frac{c}{8\pi} \int_0^{2\pi} \epsilon_i^{1/2} \, |E'(\mu')|^2 d\phi = S_i\left(\frac{\pi}{2}, \beta, z\right)\sin^2\theta' + S_i(0, \beta, z)\cos^2\theta' \quad (29)$$

where

$$S_i\left(\frac{\pi}{2}, \beta, z\right) = \frac{c\epsilon_i^{1/2}}{8} [|E'(\hat{x}')|^2 + |E'(\hat{y}')|^2] \quad (30a)$$

and

$$S_i(0, \beta, z) = \frac{c\epsilon_i^{1/2}}{4} |E'(\hat{z}')|^2 \quad (30b)$$

2. Normalization

The integrand of Equation 24d requires, not the rate of emitted energy S_i from a dipole of fixed magnitude $|\mu'|$, but the rate of emitted energy S_i *per unit of absorbed energy* (or, in quantum terms, per absorbed photon). The distinction between nonnormalized S_i and normalized $\hat{S}_i$ is crucial. A fluorophore in steady-state will dissipate its total energy (into light and heat) only at the rate it absorbs energy. A dipole of fixed magnitude $|\mu'|$, however, dissipates total energy at a rate that varies with its orientation and distance from the surface and the nature of the surface coating. This emission behavior is completely distinct from the somewhat less complicated absorption behavior as derived from the TIR evanescent wave and contained in the $|\mu \cdot E|^2$ factor. Therefore, the ϕ-integrated emission intensity S_i derived for a given dipole μ' must be normalized by the total rate P_t at which the dipole absorbs and releases energy.

We give a specific example illustrating the necessity of the normalization. Assume that we excite with p-polarized incident light at the critical angle so the evanescent wave is constant over z. For a given dipole orientation (e.g., normal to the surface), a dipole of constant magnitude μ' will absorb energy at a constant rate, independent of z. But the nonnormalized rate of emission S_i will be an exponentially decreasing function of z for any particular polar angle of observation in the glass greater than the critical angle (since the near-field of the dipole that interacts with the glass exponentially decays with z). Likewise, the total radiated power integrated over *all* propagation angles will be a function of z. But such a z-dependence of total emitted power is impossible in our particular steady-state example because energy is being absorbed in a z-independent fashion. This contradiction is resolved by altering the assumption of $|\mu'|$ constancy: in fact, $|\mu'|$ decreases somewhat as we move the dipole toward the surface. Mathematically, we need only normalize the emitted radiation energy rate derived from a constant $|\mu'|$ dipole by the total released energy rate P_t.

For nonheat dissipative surfaces such as glass, P_t is simply an integral over all radiative energy.[11] But a more general approach, applicable also to metal coatings (which dissipate some electromagnetic energy into heat), employs an expression given by Ford and Weber.[8]

$$P_t(\theta', z) = \frac{ck_1^4}{2\epsilon_1^{3/2}} \text{Re} \int_0^\infty dv \frac{v}{(1-v^2)^{1/2}} \left\{ \mu_\perp'^2 \, v^2 [1 + r_p \, e^{2ik_1 z(1-v^2)^{1/2}}] \right.$$

$$\left. + \frac{\mu_\parallel'^2}{2} [1 + r_s e^{2ik_1 z(1-v^2)^{1/2}}] + \frac{\mu_\parallel'^2}{2} (1-v^2) [1 - r_p e^{2ik_1 z(1-v^2)^{1/2}}] \right\} \quad (31)$$

where:

$$\mu_\perp'^2 \equiv \mu'^2 \cos^2\theta'$$

$$\mu_\parallel'^2 \equiv \mu'^2 \sin^2\theta'$$

The Fresnel coefficients $r_{s,p}$ are given by Equations 26 and 28. Note that $P_t(\theta', z)$ can be broken into two parts, for dipole components normal or parallel to the surface:

$$P_t(\theta', z) = P_t(0, z) \cos^2\theta' + P_t\left(\frac{\pi}{2}, z\right) \sin^2\theta' \quad (32)$$

We can now write expressions for the normalized $\hat{S}_i$ as required in Equation 24d:

$$\hat{S}_i(\theta', \beta, z) \equiv S_i(\theta', \beta, z)/P_t(\theta', z) = \frac{\hat{S}_i(0, \beta, z)}{1 + [\eta(z)\tan^2\theta']} + \frac{\hat{S}_i\left(\frac{\pi}{2}, \beta, z\right)}{1 + [\eta(z)\tan^2\theta']^{-1}} \quad (33)$$

where:

$$\eta(z) \equiv P_t\left(\frac{\pi}{2}, z\right)/P_t(0, z) \quad (34)$$

Likewise, $\hat{Q}_i(\theta', \beta_0, z)$, as an integral of $\hat{S}_i$ over the range of the objective of β, can also be expressed for arbitrary dipole orientation as the sum of two terms involving only dipoles parallel and perpendicular to the surface:

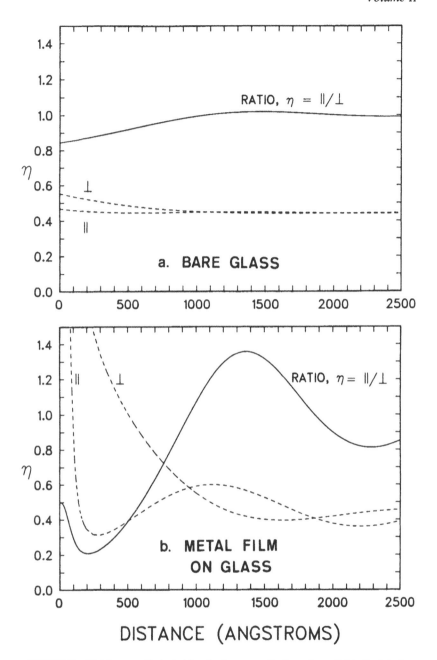

FIGURE 9. Total power dissipated by dipoles perpendicular and parallel to the interface (dashed lines) and the ratio of these powers, η (solid line) vs. dipole distance z. (a) Bare glass; (b) 20 nm aluminum film. Dielectric constants and λ_0 are the same as in Figure 8.

$$\hat{Q}_i(\theta', \beta_0, z) = \frac{\hat{Q}_i(0, \beta_0, z)}{1 + [\eta(z)\tan^2\theta']} + \frac{\hat{Q}_i\left(\frac{\pi}{2}, \beta_0, z\right)}{1 + [\eta(z)\tan^2\theta']^{-1}} \qquad (35)$$

Figures 9a and b show $P_t\left(\frac{\pi}{2}, z\right)$, $P_t(0, z)$, and $\eta(z)$ vs. z for typical experimental conditions.

Ratio $\eta(z)$ is useful because it allows a generalization of $\hat{S}$ to arbitrary dipole orientation without necessitating new integrations.

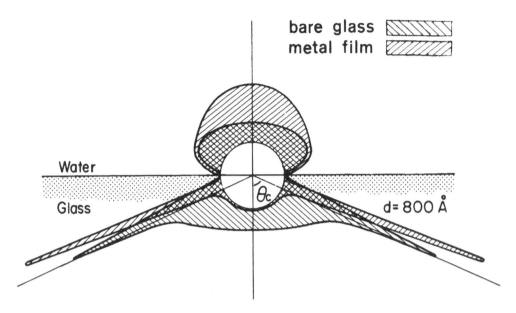

FIGURE 10. Emission intensity (observed at large distances) integrated over azimuthal observation angle ϕ vs. polar observation angle β for a layer of isotropically oriented dipoles at $z = 80$ nm from the interface. Two cases are shown; water/glass, and water/metal/glass interfaces where the layer of aluminum metal is 20 nm thick. The dielectric constants and λ_0 are as in Figure 8.

The protocol for calculating collected power P_c from Equation 24 can be summarized as follows: (a) by using Table 2 for $|\mathbf{E}'|^2$ of oriented dipoles, find $S_i\left(\frac{\pi}{2}, \beta, z\right)$ and $S_i(0, \beta, z)$ from their definitions in Equations 30a and 30b, (b) normalize these two S_i by $P_t\left(\frac{\pi}{2}, z\right)$ and $P_t(0, z)$, respectively, (c) derive the generalized $\hat{S}_i$ by Equation 33, and (d) numerically integrate Equation 24d.

C. Conclusions

Rather than displaying typical results for P_c we examine two other partly integrated measures of fluorescence from Equation 24: $F_i^{p,s}$ and $\hat{Q}$.

Figure 10 shows $F_i^p(\beta)$ on both bare and metal-coated glass for the special (but common) case where $C(\Omega', z)$ is independent of Ω' and is concentrated in a two-dimensional plane (e.g., a fluorescent-labeled cell membrane) at $z = 80$ nm. The most outstanding feature is the nonisotropic distribution of emitted radiation, which peaks at the critical angle at a bare-glass surface and at the surface plasmon resonance angle[9] at metal-coated glass.

Figure 10 suggests an elegant, but as yet untried, experimental method for selectively gathering fluorescence originating near a surface without necessitating TIR illumination. The emission power propagating into the glass at angles $\beta > \theta_c$ is substantial for dipoles near the surface (as depicted in the figure), but it rapidly approaches zero as z increases. A sufficiently high aperture objective (i.e., N.A. > 1.33) could be modified with an opaque disk in its back focal plane to block all but those rays propagating at $\beta > \theta_c$ in the glass. The objective will thereby gather light only from fluorophores near the surface, regardless of the excitation mechanism.

The ratio $\hat{Q}_i(\theta', \beta_0, z)$ is basically the probability that each excitation will result in an observed emission photon. Figures 11a, b, and c show $\hat{Q}_i(0, \beta_0, z)$ and $\hat{Q}_i\left(\frac{\pi}{2}, \beta_0, z\right)$ for

COLLECTION EFFICIENCIES

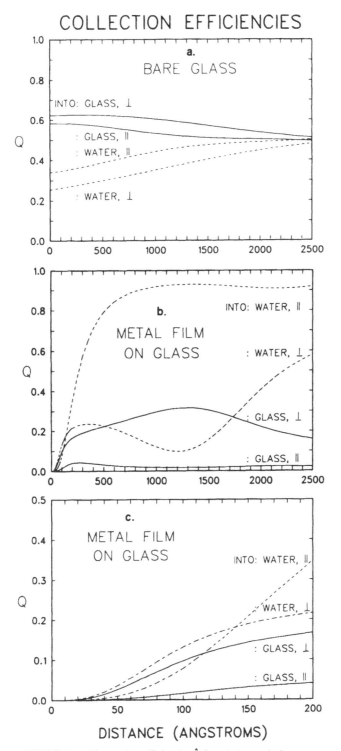

FIGURE 11. Observation efficiencies $\hat{Q}$ for a 1.4 numerical aperture objective vs. dipole distance z. Two positions of the objective are indicated; collection of light emitted into the water (dashed lines), and into the glass (solid lines). Two dipole orientations are indicated, parallel and perpendicular to the interface. Dielectric constants and λ_0 are as in Figure 8. (a) Bare glass; (b) 20 nm aluminum metal film; (c) expanded scale of (b).

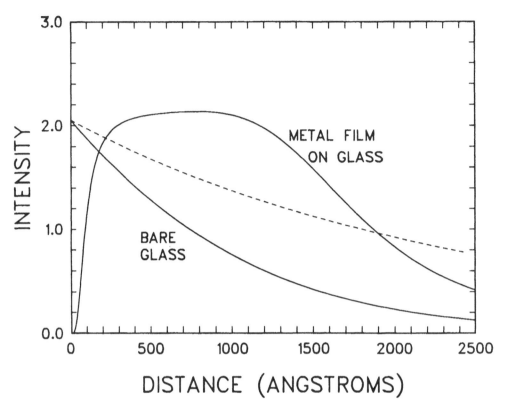

FIGURE 12.   Fluorescence intensity emitted at polar angle $\beta = 70°$ into the glass, F_3^p ($\beta = 70°$) vs. dipole distance z for dipoles oriented perpendicular to the interface. The excitation illumination is an evanescent wave (dashed line), with exponential decay $\exp(-z/250$ nm). Two cases are shown: water/glass and water/metal/glass interfaces. Dielectric constants and λ_0 are as in Figure 8. Note that although the bare glass case monotonically decreases with z, it is not an exponential decay.

some typical experimental conditions on bare and metal-coated glass. Several noteworthy features can be seen: (a) there is a very strong and highly z-dependent quenching of observed fluorescence at metal-coated surfaces (as discussed in some detail in References 8, 10, and 15). This quenching is due to the rapid increase of the total rate of released energy P_t on metal as $z \to 0$. The quenching effect may find significant applications in biophysics; e.g., distinguishing between halves of a probe-labeled lipid bilayer by selective quenching of fluorescence in the monolayer nearest an adjacent metal substrate, (b) the collected fluorescence depends on the orientation of the dipoles and the nature of this dependence is different for viewing through the glass or water side of the surface, and (c) even on bare glass, $\hat{Q}_i$ is a nontrivial function of z, again emphasizing that the intensity of observed fluorescence excited by TIR does *not* decay exponentially with distance of the dipole from the surface. Figure 12 illustrates this point by showing the observed fluorescence F_p from dipoles oriented perpendicular to the surface as a function of their distance z from a bare or metal-coated glass surface, for the selected observation angle $\beta = 70°$.

V. RECENT TIRF APPLICATIONS TO BIOSURFACES

This section is a review of recent applications of TIRF on biosurfaces.

A. Protein Adsorption to Artificial Surfaces
TIRF has been used to study equilibrium adsorption of proteins to artificial surfaces, both

to learn about the surface properties of various biomaterials that have medical applications, and also to test the TIRF technique itself.

Several studies of the binding equilibria, kinetics, and conformational changes of plasma proteins upon adsorption have employed extrinsic fluorophores (usually fluorescein, rhodamine, NBD, or dansyl) covalently attached to the protein.[1] It is possible in principle that such extrinsic groups might themselves affect the adsorption process being investigated. To avoid this possibility, the intrinsic fluorescence of tryptophan residues in the protein can be monitored upon excitation by a $\lambda_0 = 280$-nm evanescent wave.[16] In certain cases, however, the greater susceptibility of proteins to photodegradation under UV illumination may outweigh the natural advantage of intrinsic fluorescence studies.

Hlady et al.[17] propose a method for the approximate calibration of TIR fluorescence to obtain absolute surface concentrations of adsorbed protein. This method, involving use of nonadsorbing species as standards and ^{125}I-labeled protein with a γ-detection system, introduces a small correction for incident light scattered beyond the evanescent wave volume and for changes in the fluorescence emission quantum efficiency of proteins upon adsorption.

Preliminary steps have been taken to transform the sensing of protein adsorption by TIRF from a laboratory technique into a practical medical procedure.[18,19] In this application, a single multimode optical fiber, both supports the excitation evanescent wave on its surface, and also guides the captured near-field fluorescence (which propagates into the fiber at greater than the critical angle: see Section IV). Designed to serve as a continuous sensor element in a remote sample, the optical fiber is prepared with its end portion stripped bare of cladding and its very tip opaquely coated to prevent propagation of excitation of emission light into the liquid. To make the fiber biochemically specific, it is covalently coated with either an antibody (or its complementary protein antigen). When introduced into the target liquid, the antigen (or its complementary antibody) adsorbs. This adsorption can be detected by an increase in the intrinsic tryptophan fluorescence excited by $\lambda_0 = 280$ nm and detected at $\lambda = 340$ nm. Further experimentation will show whether the increase resulting from specific binding will be evident above a large background of nonspecifically adsorbed protein likely to be encountered in biological fluids. If the soluble protein that specifically adsorbs to the fiber can be extrinsically labeled, the background problem is avoided. It is then conceivable that a chemical competition for specific surface binding between the extrinsically labeled protein solution in a closed volume and the unlabeled protein in the biological fluid under investigation may give rise to an altered fluorescence signal.

B. Concentration and Orientation Distribution of Molecules Near Surfaces

To explore the fluorophore concentration $C(z)$ as a function of distance z from the interface, one can vary either the angle of incidence or the angle of observation of fluorescence (and thereby change the effective depth from which fluorescence is gathered). Reichert et al.[20] have tested this approach with a fluorescein solution to give a constant C independent of z, and with a layer of fluorescein-labeled immunoglobulin adsorbed to quartz to give a step function $C(z)$. Nevertheless, their theoretical expression is an approximation, since it omits the normalization step discussed in Section IV. This omission, although not exactly correct, simplifies the calculation of $C(z)$ by converting it to an inverse Laplace transform of the observed fluorescence. A similar simplifying approximation has been used by Allain et al.[21] in analyzing their experiments on a flexible fluorescent anthracene-polystyrene copolymer coil in the vicinity of a nonadsorbing wall. The analysis appears to confirm a local decrease in $C(z)$ for small z at the solid/solution interface. Such a depletion layer is interpreted in terms of an "entropic repulsion" model, whereby certain conformations of polymer are sterically prohibited near the surface. Analogous TIRF experiments on a stiff, high molecular weight polysaccharide, again indicate a surface-depletion zone.[22]

Another kind of ordering on surface-orientational anisotropy was investigated by TIRF.[23]

The polarization of the evanescent wave was used to selectively excite favorable orientations of fluorescent-labeled phosphatidylethanolamine embedded in lecithin monolayers on hydrophobic glass. The total fluorescence gathered by a high-aperture objective was then measured as a function of monolayer surface pressure. When interpreted according to the approximate (nonnormalized) theory[12] the results give a measure of the orientational order of the probe.

C. Morphology of Biological Cells in Culture

TIRF can be used to selectively illuminate cell-substrate contact regions in fluorescent-labeled living cell cultures[24] for qualitative study. Recently, Lanni et al.[25] have performed quantitative (but still approximate) calculations of the distance between fluorophores on (or in) a cell and the substrate, using TIRF intensities measured as a function of incidence angle. This effort is experimentally challenging, since interpretation of TIRF intensities depends upon accurate knowledge of the incidence angle to fractions of a degree and the angle-dependent changes of incident light intensity due to reflection and transmission coefficients at numerous optical interfaces in the apparatus. To analyze the data, the single-interface expressions for evanescent intensity were generalized to a model involving four layers (glass, culture medium, membrane, and cytoplasm) and simplifying approximations for the intensity of dipole emission at such a set of surfaces were made. For 3T3 cells, the authors thereby derive a plasma membrane/substrate spacing of 49 nm for focal contacts and 69 nm for "close" contacts elsewhere. They were also able to calculate an approximate refractive index for the cytoplasm of 1.358 to 1.374.

A variation of TIRF to observe cellular morphology, introduced by Gingell et al.,[26] produces essentially a negative of the standard fluorescence view of labeled cells. Here, the solution surrounding the cell is doped with a nonadsorbing and nonpermeable fluorescent volume marker. Focal contacts then appear as dark areas and other areas appear brighter, depending on the depth of solution illuminated by the evanescent wave in the cell/substrate gap.

D. Reaction Rates at Biosurfaces

TIRF intensity can be monitored as a function of time to measure the binding kinetics of a fluorophore-labeled species in solution with a TIR surface. After some sort of perturbation, the fluorescence relaxes toward equilibrium with an observable time course that depends on the kinetic rate.

Using a concentration jump as the perturbation, Sutherland et al.[27] measured the kinetics of binding of fluorescein-labeled human IgG (as an antigen in solution) to surface-immobilized sheep antihuman IgG. The TIRF surface was either a planar slide or a fiber-optic cylinder.

To increase the speed of TIRF-based kinetic techniques, the perturbation can be optical rather than chemical. A bright brief flash of evanescent light irreversibly photobleaches those species bound in equilibrium to the surface. The photobleaching flash is then immediately reduced in intensity several thousand-fold to avoid further bleaching. Fluorescence is then seen to "recover" as unbleached species from solution exchange with bleached species desorbing from the surface. The observed fluorescence recovery rate is a measure of the chemical kinetic rates of the adsorption/desorption process.[2,3] This technique is dubbed TIR/FPR in reference to fluorescence photobleaching recovery.

For TIR/FPR to be useful for chemical kinetics studies on biological membranes, as opposed to artificial surfaces, two problems must be confronted: (1) how to position the membrane in a TIR system; and (2) how to overcome background binding to the substrate to which the membrane is attached. We have successfully dealt with these problems in recent preliminary TIR/FPR measurements of nonspecific reversible binding kinetics of fluorescent-

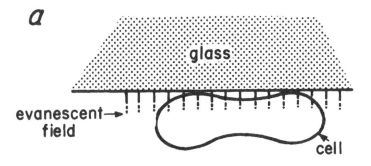

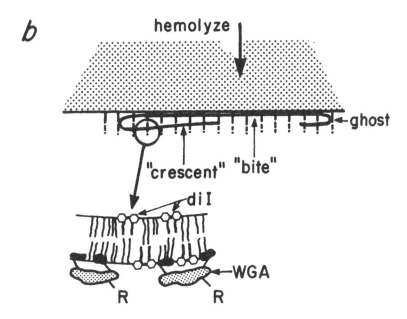

FIGURE 13. Schematic drawing of the likely morphology of spread erythrocyte ghosts. Lipid probe diI is 3,3'-dioctadecylindocarbocyanine; R-WGA is tetramethyl-rhodamine-labeled wheat germ agglutinin.

labeled epidermal growth factor (EGF) to human erythrocyte membrane.[28] Because of their potential general applicability, these experimental protocols are described here in some detail.

Human erythrocytes are adsorbed to a 1 in. × 1 in. surface of either bare or aluminum vacuum-coated microscope slide glass that has been covalently bonded with poly-*l*-lysine. The procedure for covalent attaching poly-*l*-lysine, which involves presilanization and suc-cinylation of glass has been described for treatment of spherical glass beads.[29]

After the red cells are adsorbed onto the surface, they are hemolyzed by hypoosmotic shock. The ghost membrane apparently spread out as depicted in the schematic drawing of Figure 13. This morphology is deduced from the characteristic "crescent" and "bite"-labeling patterns of lipid probe diI, rhodamine wheat germ agglutinin (R-WGA), and antispectrin.

DiI treatment, either before (prelabel) or after (postlabel) sticking and spreading of the erythrocyte to the surface, causes both the bite and the crescent to be labeled (Figure 14a). The crescent is approximately twice as bright as the bite (as viewed under epi-illumination), suggesting a single membrane layer in the bite and a double layer in the crescent.

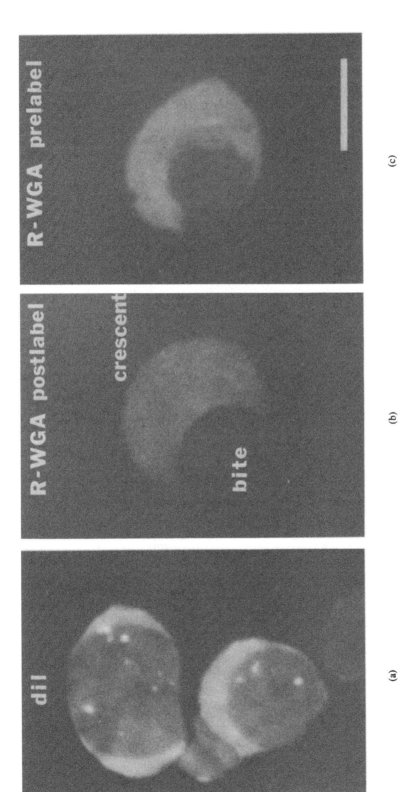

FIGURE 14. Erythrocyte ghosts spread on bare glass covalently coated with poly-*l*-lysine. (a) diI-labeled; (b) R-WGA labeled after spreading (postlabel); (c) R-WGA labeled before spreading (prelabel). Epi-illumination with 100 ×, 1.2 N.A. water-immersion objective. Space bar = 5 μm.

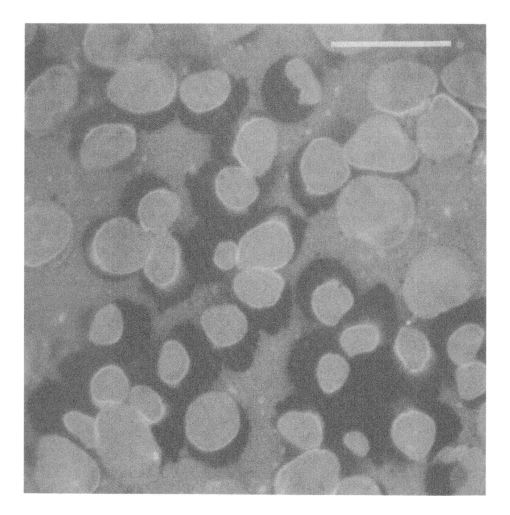

FIGURE 15. Spread erythrocytes treated with sheep antihuman spectrin antibodies (a gift of Jon Aster and George Brewer) and then labeled with fluorescein antisheep antibodies. Epi-illumination; objective was 50 ×, N.A. = 1.00, water immersion. Space bar = 20 μm.

With R-WGA postlabeling (Figure 14b) only the crescent is labeled. This suggests that (1) the exposed surface of the bite membrane is cytoplasmic with no WGA receptors and (2) the spacing between ghost and the solid substrate is so close that the R-WGA molecules cannot diffuse into it. With R-WGA prelabeling (Figure 14c), the bite is dimly labeled, strengthening the view that the bite consists of a single membrane with its outside surface facing the substrate and cytoplasmic surface facing the open solution. Under TIR illumination, the patterns are qualitatively similar to epi-illumination for both diI and R-WGA-labeling, suggesting that the membrane is indeed flattened on the glass.

Further confirmation that the "bite" presents the cytoplasmic surface to the solution is the selective binding of labeled antispectrin antibodies in the "bite" (Figure 15). This figure also demonstrates that the spacing between membranes in the overleafed region is too narrow for antibody molecules to diffuse in.

To study the interaction of EGF with spread ghost membranes, the microscope slide with adsorbed membranes is placed in equilibrium with a solution of fluorescein-EGF in Hank's balanced salt solution without calcium or magnesium (HBSS/Ca-free) in a chamber similar to that in Figure 1. When viewed with epi-illumination, the fluorescence of F-EGF in bulk solution totally obscures the fluorescence of F-EGF adsorbed to either the ghost membranes or surrounding substrate. But when viewed by TIR, fluorescence from the surface dominates. Unfortunately, however, the concentration of F-EGF adsorption directly onto a polylysinated glass substrate is very large compared to that on the ghost membranes and the fluorescent field again appears uniform at least initially. But as fluorescein-EGF bleaches during TIRF observation, the narrow aqueous layer between a ghost membrane and the substrate becomes dark due to a slow diffusional exchange of bleached and unbleached F-EGF. Each membrane, with its characteristic crescent and bite, then becomes discernible as a dark disk on a bright background (Figure 16a).

This overwhelming "background" fluorescence of F-EGF adsorbed to the substrate can be overcome by utilizing the quenching effect of a metal surface on a nearby fluorophore, as discussed theoretically in Section IV. We first vacuum coat the glass with ~200 Å of aluminum, expose it to air to produce an oxide layer, then subsequently silanate, succinylate, and poly-*l*-lysinate the surface, and finally adsorb and hemolyze erythrocytes in exactly the same manner as on bare glass. When placed in equilibrium with F-EGF and illuminated by TIRF, the ghost membranes on the aluminum coating stand out as bright disks on a dark background (Figure 16b). Clearly, fluorescence from F-EGF adsorbed to the membrane in the bite and crescent regions is not quenched nearly as strongly as that adsorbed directly onto the derivatized aluminum oxide coating.

Preliminary experiments on this membrane preparation with TIR/FPR indicate that the average residence time of F-EGF on the erythrocyte ghost membrane is about 0.5 sec and that almost all the adsorption is reversible.[28]

One biophysical rationale for measuring the surface residency time of hormone nonspecifically adsorbed to a membrane is to assess whether such adsorption might affect the reaction rate with specific hormone receptors. In principle, nonspecific adsorption followed by two-dimensional surface diffusion can enhance specific reaction rates under certain conditions.[28,30,31] Whether such enhancement actually occurs depends on the surface-diffusion coefficient. TIR/FPR can measure the rate of surface diffusion if one introduces some sort of small-scale spatial variation in the evanescent wave pattern for both photobleaching and excitation during the recovery phase. One such spatial variation is simply a focused line, as employed by Burghardt and Axelrod.[3] Another is a series of parallel lines, as depicted in Figure 5, created by intersecting TIR beams.

E. Cytoplasmic Filament Distribution

Cytoplasmic filaments can be visualized by a number of fluorescence-tagged agents including antibodies, toxins, and fluorescent analogs of filament forming or binding proteins. On thick cells, however, out-of-focus fluorescence excited by epi-illumination obscures details in the plane of focus. By exciting a very thin layer in the cell near substrate contact points, TIRF can clearly visualize filaments in these regions otherwise difficult to see.

In collaboration with Dr. Robert Bloch of the University of Maryland and Marisela Velez of the University of Michigan, we have used TIRF to study cytoplasmic filament distribution in the vicinity of acetylcholine receptor (AChR) clusters on developing muscle cells (rat myotubes) in culture. TIRF, interference reflection contrast microscopy, and other techniques have all shown that these highly punctate clusters on rat myotubes occur mainly in regions of close cell-substrate contact. Because of the prominence of such AChR clusters, both on nerve-free rat myotube cultures and also at synapses in vivo, much attention has been devoted to finding what triggers AChR to cluster and what molecules anchor the clusters. The answers are not yet clear, but certain labeled cytoplasmic filamentous structures do exhibit, under TIRF illumination, a complex pattern in the region of the AChR clusters. For example, Figure 17 shows that the cytoplasmic protein vinculin (marked by rhodamine antibodies) and AChR (marked by fluorescein α-bungarotoxin) are anticorrelated in their distribution;

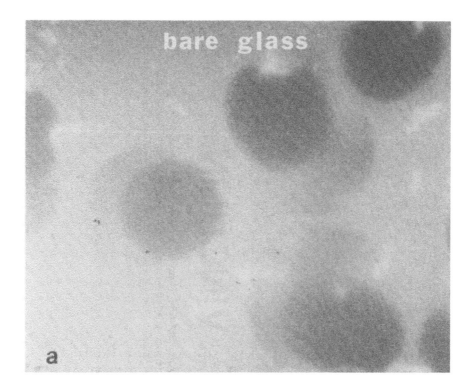

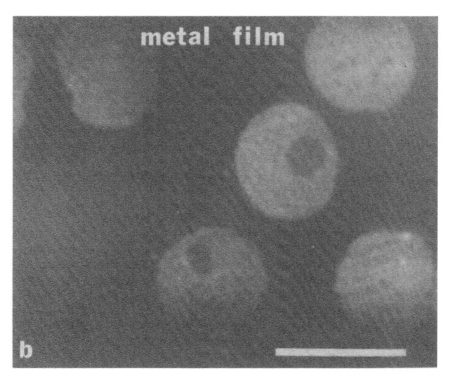

FIGURE 16.   Spread erythrocytes in equilibrium with 30 μg/mℓ fluorescein-EGF in Hank's balanced salt solution (Ca-free), illuminated by TIRF with λ_a = 476.5 nm on poly-*l*-lysine coated (a) bare glass or (b) aluminum film (200 Å-thick) coated glass. Objective was 100 ×, N.A. = 1.20, water immersion. The pattern on bare glass becomes visible only after some photobleaching occurs. Space bar = 10 μm.

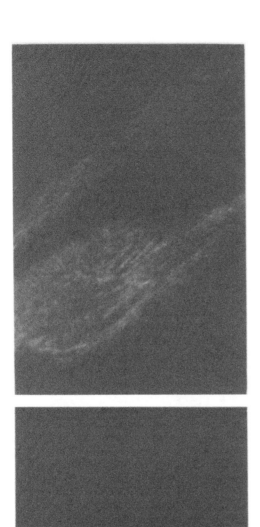

AChR

VINCULIN

FIGURE 17. TIRF photographs at AChR clusters on rat myotubes in primary culture of AChR (labeled by fluorescein α-bungarotoxin) and vinculin (indirect, labeled by rhodamine antibodies). Rabbit antivinculin IgG was a gift of Robert Bloch; the cell preparation and photographs were by Marisela Velez. Two different scenes are shown; both illustrate the fine-scale anticorrelation in the distributions of AChR and vinculin. Objective was 100 ×, N.A. = 1.20, water immersion. Space bar = 20 μm.

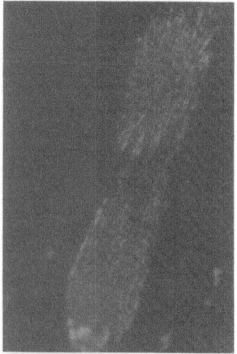

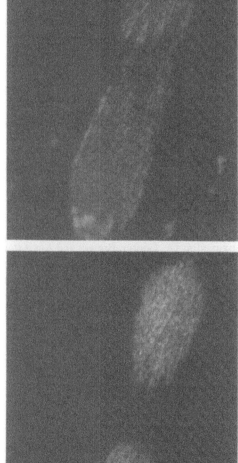

VINCULIN

AChR

FIGURE 17. Continued.

i.e., where AChR are particularly concentrated in speckles and streaks within a cluster, vinculin is scarce, and vice versa. We expect that the actual molecular anchor will exhibit a distribution that positively correlates with AChR, and TIRF will thereby aid in identifying such an anchor.

VI. SUMMARY

TIRF is an experimentally simple technique for selective excitation of fluorophores on or near a solid surface. Applications in biology, biochemistry, and biophysics are still evolving and include:

1. Localization of cell-substrate contact regions in cell culture and determination of the distribution of specific cell-surface components relative to those contact regions
2. Fluorescence excitation of cytoplasmic filaments near the membrane at substrate contacts, while avoiding excitation of fluorescence from other planes of focus
3   Study of reversible adsorption kinetics and surface diffusion on biosurfaces, including biological membranes
4. Measurement of concentration and orientational distribution of adsorbants as a function of distance from the surface
5. In vivo medical sensors based on fiber optics
6. For TIRF on metal-coated surfaces in contact with a membrane, possible measurement of transmembrane molecular dynamics

ACKNOWLEDGMENTS

We thank Dr. Robert Bloch for his gift of antivinculin; Dr. George Brewer and Jon Aster for their gift of antispectrin; Yvette Dorsey and Marisela Velez for their preparation and photography of rat myotubes; Dr. Robert Petrisko for advice concerning the derivatization of aluminized surfaces, and Drs. Nancy Thompson, G. W. Ford, and W. H. Weber for helpful discussions concerning fluorophore emission at surfaces. This work was supported by PHS-NIH grant NS 146565.

REFERENCES

1. **Axelrod, D., Burghardt, T. P., and Thompson, N. L.,** Total internal reflection fluorescence, *Annu. Rev. Biophys. Bioeng.*, 13, 247, 1984.
2. **Thompson, N. L., Burghardt, T. P., and Axelrod, D.,** Measuring surface dynamics of biomolecules by total internal reflection fluorescence with photobleaching recovery or correlation spectroscopy, *Biophys. J.*, 33, 435, 1980.
3. **Burghardt, T. P. and Axelrod, D.,** Total internal reflection/fluorescence photobleaching recovery study of serum albumin adsorption dynamics, *Biophys. J.*, 33, 455, 1981.
4. **Weis, R. M., Balakrishnan, K., Smith, B. A., and McConnell, H. M.,** Stimulation of fluorescein in a small contact region between rat basophil leukemia cells and planar lipid membrane targets by coherent evanescent radiation, *J. Biol. Chem.*, 257, 6440, 1982.
5. **Burghardt, T. P. and Thompson, N. L.,** Evanescent intensity of a focused Gaussian light beam undergoing total internal reflection in a prism, *Opt. Eng.*, 23(1), 062, 1984.
6. **Born, M. and Wolf, E.,** *Principles of Optics,* 5th ed., Pergamon Press, Oxford, 1975.
7. **Simon, H. J., Mitchell, D. E., and Watson, J. G.,** Surface plasmons in silver film. A novel undergraduate experiment, *Am. J. of Phys.*, 43(7), 630, 1975.
8. **Ford, G. W. and Weber, W. H.,** Electromagnetic interactions of molecules with metal surfaces, *Phys. Rep.*, 113(4), 195, 1984.

9. **Weber, W. H. and Eagen, C. F.**, Energy transfer from an excited dye molecule to the surface plasmons of an adjacent metal, *Opt. Lett.*, 4, 236, 1979.
10. **Chance, R. R., Prock, A., and Silbey, R.**, Molecular fluorescence and energy transfer near interfaces, *Adv. Chem. Phys.*, 37, 1, 1978.
11. **Lukosz, W. and Kunz, R. E.**, Light emission by magnetic and electric dipoles close to a plane interface. I. Total radiated power, *J. Opt. Soc. Am.*, 67(12), 1607, 1977.
12. **Burghardt, T. P. and Thompson, N. L.**, Effect of planar dielectric interfaces on fluorescence emission and detection: evanescent excitation with high aperture collection, *Biophys. J.*, 46, 729, 1984.
13. **Carniglia, C. K., Mandel, L., and Drexhage, K. H.**, Adsorption and emission of evanescent photons, *J. Opt. Soc. of Am.*, 62(4), 479, 1972.
14. **Lee, E.-H., Benner, R. E., Fen, J. B., and Chang, R. K.**, Angular distribution of fluorescence from liquids and monodispersed spheres by evanescent wave excitation, *Appl. Opt.*, 18(6), 862, 1979.
15. **Drexhage, K. H.**, Interaction of light with monomolecular dye layers, in *Progress in Optics*, Vol. 12, Wolf, E., Ed., North-Holland, Amsterdam, 1974, 165.
16. **Hlady, V., Van Wagenen, R. A., and Andrade, J. B.**, Total internal reflection intrinsic fluorescence (TIRIF) spectroscopy applied to protein adsorption, in *Surface and Interfacial Aspects of Biomedical Polymers*, Vol. 2, Andrade, J. D., Ed., Plenum Press, New York, 1985.
17. **Hlady, V., Reinecke, D. R., and Andrade, J. D.**, Fluorescence of adsorbed protein layers: quantitation of total internal reflection fluorescence, *J. Colloid Interface Sci.*, 3, 555, 1986.
18. **Newby, K., Reichert, W. M., Andrade, J. D., and Benner, R. E.**, Remote spectroscopic sensing of chemical adsorption using a single multimode optical fiber, *Appl. Opt.*, 23(11), 1812, 1984.
19. **Andrade, J. D., Reichert, W. M., Gregonis, D. E., and Wagenen, R. A.**, Remote fiber-optic biosensors based on evanescent-excited fluoro-immunoassay: concept and progress, *IEEE Trans. Electron Devices*, ED-32, 1175, 1985.
20. **Reichert, W. M., Suci, P. A., Ives, J. T., and Andrade, J. D.**, Evanescent detection of adsorbed protein concentration-distance profiles: fit of simple model to variable-angle total internal reflection fluorescence data, *Appl. Spectros.*, 41, 503, 1987.
21. **Allain, C., Aussere, D., and Rondelez, F.**, Direct optical observation of interfacial depletion layers in polymer solutions, *Phys. Rev. Lett.*, 49, 1694, 1982.
22. **Aussere, D., Hervet, H., and Rondelez, F.**, Concentration profile of polymer solutions near a solid wall, *Phys. Rev. Lett.*, 54, 1948, 1985.
23. **Thompson, N. L., McConnell, H. M., and Burghardt, T. P.**, Order in supported phospholipid monolayers detected by the dichroism of fluorescence excited with polarized evanescent illumination, *Biophys. J.*, 46, 739, 1984.
24. **Axelrod, D.**, Cell-substrate contacts illuminated by total internal reflection fluorescence, *J. Cell Biol.*, 89, 141, 1981.
25. **Lanni, F., Waggoner, A. S., and Taylor, D. L.**, Structural organization of interphase 3T3 fibroblasts studied by total internal reflection fluorescence microscopy, *J. Cell Biol.*, 100, 1091, 1985.
26. **Gingell, D., Todd, I., and Bailey, J.**, Topography of cell-glass apposition revealed by total internal reflection fluorescence of volume markers, *J. Cell Biol.*, 100, 1334, 1985.
27. **Sutherland, R. M., Dahne, C., Place, J. F., and Ringrose, A. S.**, Optical detection of antibody-antigen reactions at a glass-liquid interface, *Clin. Chem. (Winston-Salem, NC)*, 30, 1533, 1984.
28. **Axelrod, D., Fulbright, R. M., and Hellen, E. H.**, Adsorption kinetics on biological membranes: measurement by total internal reflection fluorescence, in *Fluorescence in the Biological Sciences*, Taylor, D. L., Waggoner, A. S., Lanni, F., Murphy, R. F., and Birge, R., Eds., Alan R. Liss, New York, 1985.
29. **Jacobson, B. S., Cronin, J., and Branton, D.**, Coupling polylysine to glass beads for plasma membrane isolation, *Biochim. Biophys. Acta*, 506, 81, 1978.
30. **Adam, G. and Delbruck, M.**, Reduction of dimensionality in biological diffusion processes, in *Structural Chemistry and Molecular Biology*, Rich, A. and Davidson, N., Eds., W. H. Freeman, San Francisco, 1968, 198.
31. **Berg, H. and Purcell, E. M.**, Physics of chemoreception, *Biophys. J.*, 20, 193, 1977.
32. **Hellen, E. H. and Axelrod, D.**, Fluorescence emission at dielectric and metal-film interfaces, *J. Opt. Soc. Am. (B)*, 4, 337, 1987.
33. **Thompson, N. L. and Burghardt, T. P.**, Total internal reflection fluorescence: measurement of spatial and orientational distributions of fluorophores near planar dielectric interfaces, *Biophys. Chem.*, 25, 91, 1986.

Chapter 12

NEAR-FIELD IMAGING OF FLUORESCENCE

Aaron Lewis, Eric Betzig, Alec Harootunian, Michael Isaacson, and Ernst Kratschmer

TABLE OF CONTENTS

I. INTRODUCTION

Fluorescence microscopy and imaging discussed in the other chapters of this book are based on the principles of geometrical optics and are fundamentally limited in their resolution by the wavelength of light. This is an inviolate physical characteristic of such methods, in spite of the recent introduction of image-enhancement techniques. However, we have developed a new form of fluorescence microscopy and imaging that has a resolution that is independent of wavelength.[1-6] These new fluorescence methods do not use lenses which focus light to a diffraction-limited spot but, rather, employ submicron apertures which can be used to form a subwavelength light beam. The resulting techniques are called near-field scanning optical microscopy (NSOM) and near-field statistical analysis of submicron structures (NSASS). In this chapter we describe the physical basis of this new methodology, the experimental methods that have been used and are being developed to translate these physical principles into experimental reality, the signal-to-noise considerations, and the potential for this new form of microscopy and spectral imaging in biophysical research.

II. THE PHYSICAL PRINCIPLES

In order to understand fluorescence NSOM or NSASS it is necessary to consider the spatial distribution of electromagnetic radiation as it emanates from a submicron aperture. Betzig et al.[6,7] have considered this problem from a theoretical point of view. These studies have demonstrated that the radiation emanating from an aperture is at first highly collimated to the aperture dimension. In this region, called the near-field (see Figure 1), a submicron light beam exists that has a dimension solely dependent on the aperture diameter and independent of the wavelength. Subsequent to traversing this near-field regime the collimated light beam spreads into a diffraction-limited spot of light in the far-field (see Figure 1). All the fluorescence microscopy and imaging reported to date have been performed in this far-field regime where geometric optics define the resolution that can be achieved with lenses. Under such conditions, details cannot be resolved to better than $\simeq \lambda/2$, where λ is the wavelength of the fluorescence, and even this limit is not achieved often in fluorescence microscopy.

Operating in the near-field allows us to overcome these limitations of geometric optics. If an illuminated submicron aperture is brought close ($\simeq 250$ Å for a 500 Å aperture[6,7]) to a surface, such as a cell membrane embedded with dye molecules that can fluoresce, then the beam emanating from the aperture excites a submicron region of the surface. The fluorescent molecules in the excited region act as donors and the aperture then acts as an acceptor to efficiently receive and transmit the fluorescent light through the aperture. The transmitted light can then be collected with conventional optical elements. The distances over which such efficient energy transfer can occur have been defined in many experiments.

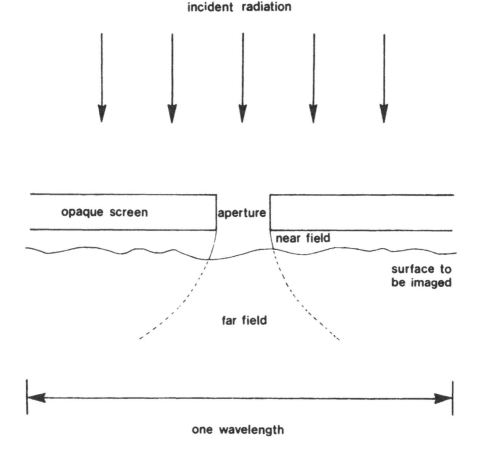

incident radiation

opaque screen | aperture

near field

surface to
be imaged

far field

one wavelength

FIGURE 1.   A diagramatic representation of the collimation in the near-field of radiation emanating from a subwavelength aperture. Notice that as a function of distance from the aperture the radiation field spreads out. The regime in which the radiation has spread out is known as the far-field. All the microscopy and imaging reported in the other chapters of this book are constrained by the limits of the far-field. The far-field limit of resolution is approximately 1/2 of the wavelength and this limit is not achieved often in fluorescence microscopy. In near-field microscopy the resolution is dependent not on wavelength but on the aperture diameter.

The most applicable for our purposes is the work by Kuhn, Drexhage, and co-workers.[8-11] These experiments employ Langmuir-Blodgett methods for forming monolayer and multi-layer films of fluorescent molecules. The distances of these films can be adjusted to better than 20 Å relative to metal films, mirrors, or other monolayers of dye molecules that can reflect or accept the fluorescence. Considering that the metal apertures used act as dipole acceptors of electromagnetic radiation, it can be deduced[10] from the Langmuir-Blodgett studies that the regime of efficient energy transfer can extend for as far as $\simeq 300$ Å for the very bright fluorophores available today with absorptions of 10^{-1} and quantum yields approaching unity. It is important to note that these distances are valid only for appropriate orientations of the donor relative to an electric-dipole acceptor. It should also be noted that such dipole energy transfer is a worst case. Under the appropriate conditions, where the donor is an electric-quadrupole emitter, the situation could become even better with efficient energy transfer distances approaching $\simeq 600$ Å. Practically speaking, however, since the availability of high-brightness quadrupole emitters is presently limited, our experiments are designed to be consistent with the more conservative distances associated with electric-dipole/electric-dipole energy transfer. Thus, the object of fluorescence near-field microscopy and imaging is to develop the experimental methodology to analyze the emitted light in the near-field, or in regions close to the near-field, where a spatial correlation can still be made between the emitted light and the location in the xy plane of the fluorescent molecule.

In terms of NSOM it is easy to envision the experimental design of such a near-field fluorescence microscope. It involves taking an aperture of a diameter approximately equivalent to the resolution required, bringing it close, ≈ 300 Å, to a fluorescing surface that is to be imaged and then recording the intensity of fluorescence as the aperture is scanned over the surface. In practice, however, in order to achieve such a goal, the cutting edge of many technological areas has to be synthesized. These areas include: microfabrication, micromovement, computers, lasers, and low-light level detection. The result of such a program is the development of wavelength-independent fluorescence imaging, with resolution comparable to that of a scanning electron microscope. A variety of important problems at this resolution could then be investigated in living cells and macromolecular assemblies without the need for vacuum conditions and without the use of potentially damaging ionizing radiation. Examples of problems in this resolution regime that could be probed using fluorescence NSOM include: membrane microdomain structure and receptor cluster assembly, membrane cytoskeletal interactions, organelle movement along dissociated actin filaments, and even kinetics of specific protein movements in macromolecular assemblies, as a result of a variety of chemical and other stimulants. Furthermore, a by-product of fluorescence NSOM development would be the capability of monitoring z-direction topographical alterations with a resolution of an order of magnitude or two better than our present goal of 500-Å lateral resolution. Cell membranes are known to alter their surface topography as a result of stimulation[12] and, thus, NSOM should be an extremely precise *in situ* monitor of these topographical changes. NSOM and fluorescence NSOM are important new probes in cell biology, which should make significant contributions to a wide variety of biological problems. The potential of these techniques is not limited, however, to biology. Their applicability should also extend from microelectronics to physics, chemistry, and material science, where nondestructive high-resolution imaging is required.

III. FLUORESCENCE IS TRANSMITTED THROUGH < λ/10 APERTURES

The first step in developing near-field fluorescence imaging is to construct apertures considerably smaller than the wavelength of the emission ($< \lambda/10$) and to demonstrate efficient transmission of fluorescence through such apertures. Therefore, in this section, we first describe one of the methods used in constructing well-characterized apertures and then we demonstrate that such apertures, illuminated with a variety of sources, including fluorescent molecules, do efficiently transmit visible radiation.

A. Fabrication of Aperture Arrays

Over the past few years a variety of techniques have been developed to fabricate submicron apertures and arrays of submicron apertures. Some of these methods are quite simple and others rather complex. In this section, one method for the fabrication of aperture arrays will be outlined. Although these arrays are not used in near-field microscopy, they are effectively employed in problems that can be probed with NSASS. In addition, a significant fraction of the preliminary fluorescence measurements were completed with such aperture arrays.

The procedure used for the fabrication of aperture arrays is schematically represented in Figure 2. The aperture fabrication begins with a 3-in. diameter, 250-μm thick wafer which is a single crystal of Si with a <100> orientation and polished on both sides. First, 150 nm Si_3N_4 films have to be deposited by low-pressure chemical vapor deposition (LPCVD) on both sides of the 250-μm-thick Si single crystal for later processing steps (see Figure

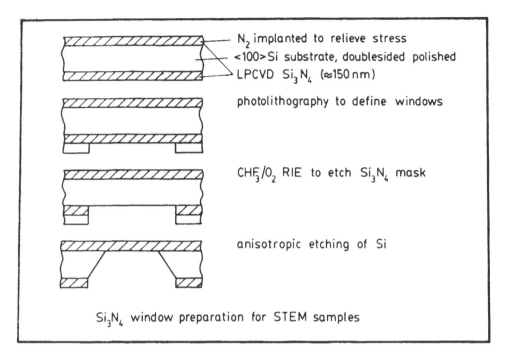

FIGURE 2.   Steps involved in the processing of 3-in.-diameter single-crystal Si wafers to produce windows in which apertures and other microfabricated structures can be enscribed.

2). For these later steps it is also necessary to release some of the tensile stress in the one Si_3N_4 film in which the apertures will finally be formed. This is done by N_2 implantation and prevents cracking of the Si_3N_4 film in the final etching procedure which forms the apertures.

With these preliminary steps completed, the Si_3N_4 film that was not implanted with N_2 is coated with photoresist, a polymer that can be photoaltered. A mask containing open squares is placed in contact with the resist and light is then shone on the photoresist through the mask. Chemicals are subsequently used to develop (remove) the photoresist in the exposed-square regions. The Si_3N_4 in these exposed-square regions is removed using reactive ion etching (RIE) with CHF_3/O_2 forming the large windows seen in the micrographs reproduced in Figure 3. Further etching (removal) with catechol:ethylenediamine:H_2O etches the Si single crystal at different rates along different crystalline axes. This process, called anisotropic etching, produces the structure in the Si seen in the last diagram of Figure 2,[13] and creates the four smaller windows (seen in Figure 3) in which the apertures are finally formed.

To form the apertures in one of the four smallest windows seen in Figure 3, the N_2-implanted Si_3N_4 layers are thinned to 50 to 100 nm and coated with a 100-nm-thick film of polymethylmethacrylate (PMMA) (see Figure 4). The electron beam of a scanning transmission electron microscope (STEM) is then employed to break up the bonds of the PMMA polymer in a predetermined pattern (see Figure 4). These exposed regions of the PMMA are then developed out with chemicals that are sensitive to the altered chemical nature of the electron beam-exposed polymer. Holes as small as 8 nm have been effectively produced using this procedure.[14] RIE is then used as before to etch the Si_3N_4. This leaves an aperture in the Si_3N_4, which is coated with metal over the PMMA as shown in the last diagram of Figure 4.

B. Light Transmission

The procedure outlined above is the method most recently used to create well-defined

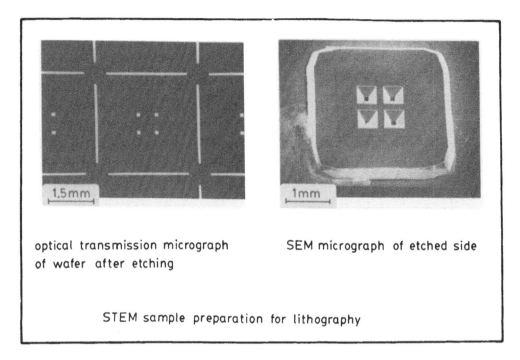

optical transmission micrograph
of wafer after etching

SEM micrograph of etched side

STEM sample preparation for lithography

FIGURE 3.   Optical and scanning electron micrographs of the larger and smaller windows enscribed in the silicon wafer. In the scanning electron microscope (SEM) micrograph the effect of the anisotropic etching is clearly visible in the four smaller windows.

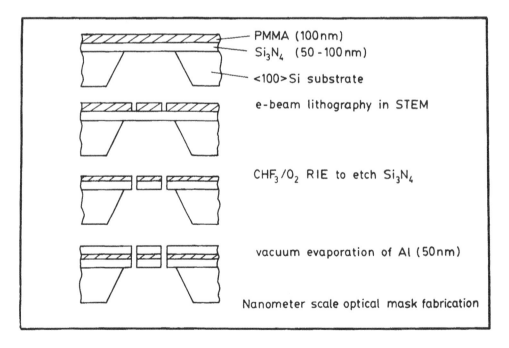

PMMA (100 nm)
Si_3N_4 (50 - 100 nm)
<100> Si substrate

e-beam lithography in STEM

CHF_3/O_2 RIE to etch Si_3N_4

vacuum evaporation of Al (50 nm)

Nanometer scale optical mask fabrication

FIGURE 4.   Procedure used to inscribe apertures on the Si_3N_4 windows.

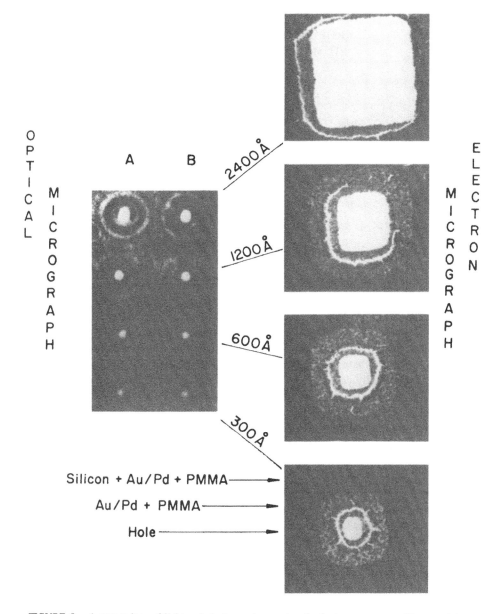

FIGURE 5.   A comparison of light and electron micrographs of submicron apertures. The material surrounding the apertures is indicated in the bottom of the figure. The light area immediately around the hole is Au/Pd and PMMA. This area is surrounded by a dark region where Si, Au/Pd, and PMMA are all present.

apertures in thin-metal films. In an early report we used a related procedure to form apertures as small as 300 Å in silicon membranes through which light transmission could be readily detected.[2] The comparison of electron and light micrographs shown in Figure 5 is taken from this first report. In this study, each aperture in the array was first analyzed using the STEM to accurately determine its geometry and dimensions. Photomicrographs were taken of these same apertures using tungsten illumination filtered to pass 5500 Å light. To obtain the photomicrographs in Figure 5, ASA 100 film was used, with few second exposures.

For these early experiments, the membrane was coated with 300 Å Au/Pd. Although this alloy has a very small grain size, its optical qualities are not as good as those for other metals, such as aluminum or chrome. Thus, there was as much as 10% transmission through

Table 1
PROPERTIES OF THIN-METAL
FILMS USED IN APERTURE
FABRICATION

Metal	Fraction of 5000 Å light transmitted through a 500 Å thick-metal film[a]	Grain size (Å)
Al	4.8×10^{-4}	100
Cr	3.8×10^{-3}	10 to 20
Pt	1.3×10^{-2}	<10
W	3.4×10^{-2}	<10
Au	1.3×10^{-1}	100

[a]   Weaver, J. W., Krafka, C., Lynch, D. W., and
Koch, E. E., Optical properties of metals, *Phys.
Dat.*, 18, 1981.

the metal, making for low contrast between the metal and the hole. In spite of these limitations and the use of a 63 × objective, images of light emanating through such small apertures could be readily recorded. We have used densitometry to quantify the light transmission through the apertures to the known transmission through the metal, and our results indicate that the transmission is greater than predicted by theoretical calculations for a subwavelength aperture in a perfectly conducting screen.[15] This may be expected because the theory is for an infinitely thin perfectly conducting screen, while the apertures used in the above experiments were in a screen that was not perfectly conducting and not infinitely thin.

The apertures that we currently use are coated with aluminum. As seen in Table 1 a 500-Å-thick film of aluminum transmits ≃ three orders of magnitude less visible light than a gold film of the same thickness. (Au/Pd films exhibit optical transmission properties similar to Au.) The degree of opacity of the film is important in order to maximize light transmission through the aperture while giving high contrast between the hole and surrounding metal film. In fact, it is this contrast that may eventually define the resolution limit of near-field microscopy and imaging.

In Figure 6 we compare the transmission for arrays of 2400, 1200, and 600-Å aluminum apertures. These records were obtained directly from the output of an optical multichannel analyzer which detected the image from the entire field of apertures in the optical microscope. The throughput through each of the apertures was simultaneously recorded and digitized. The three-dimensional graphs seen in Figure 6 are these digitized records plotted with the same scale for all the aperture arrays. Notice the excellent signal/noise of even the smallest apertures and the reproducibility in the aperture construction as evidenced by the similar light throughput for different apertures of the same size. Presently, a detailed wavelength dependence of the aperture transmission and aperture output is underway to understand the underlying physics in the modes of transmission through such apertures.

C. Fluorescence Transmission

Cellular membranes are low in contrast and although near-field imaging does not depend on the use of fluorescence, the use of fluorescence will be important to the wide applicability of near-field imaging in biology. Thus, a primary concern has been to develop the techniques that would allow the viewing of fluorescence through subwavelength apertures with high signal/noise ratio before fluorophore bleaching. For such experiments a laser was used, but the coherence of the laser makes it impossible to view aperture arrays, even in the fluorescence mode, without the elimination of coherent effects in the incident beam. Therefore, for our

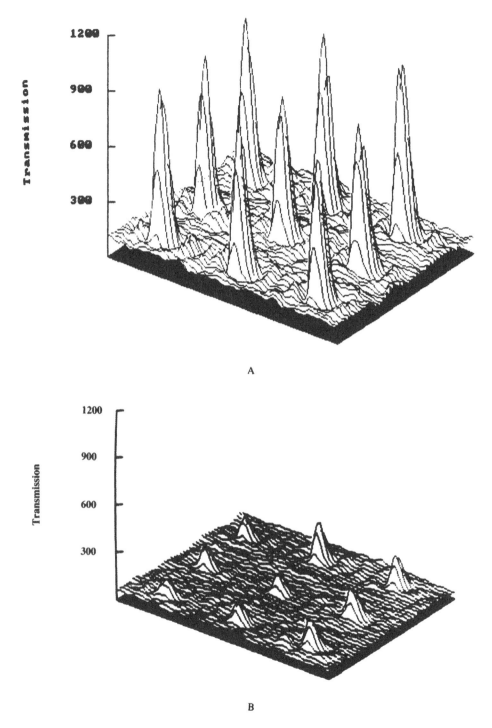

FIGURE 6. 5145-Å light transmission through (A) 2400-Å-, (B) 1200-Å-, and (C) 600-Å-diameter apertures is shown detected in the far-field with an optical multichannel analyzer. Incident beam power was 0.62 μW/cm².

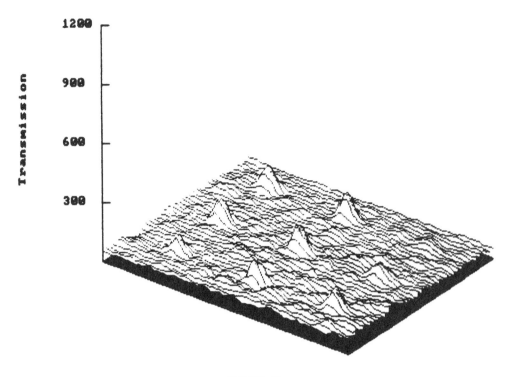

FIGURE 6C.

fluorescence measurements, in addition to the use of a laser, a spinning ground-glass slide is placed in the path of the incident beam. The fluorescence can then be viewed either in transmission or in epi-illumination. For the initial fluorescence measurements we used a thin film of perylene on a glass slide. The emission maximum was at 6900 Å.

The fluorescence pictures seen in Figure 7 were taken through the apertures with the perylene-coated slide illuminated in transmission. The largest apertures in each of three arrays were 2400, 1200 and 600 Å. Each array was constructed with electron beam intensities that were varied from high to low from left to right. Therefore, the apertures on the left were the largest and the apertures to the right smaller. The largest apertures in Figures 7A, B, and C were 2400, 1200, and 600 Å, respectively. As can be seen, fluorescence is readily detected through even the smallest apertures in the 600 Å array. The fluorescence micrographs were recorded with ASA 400 film and with no detectable bleaching. This result is a significant step forward in the application of near-field imaging to biologically significant problems.

IV. NSASS

As described in the introduction, one method of near-field imaging is to scan an aperture in near-field over a fluorescing surface and to monitor the fluorescence intensity at each aperture position. Even though the aperture arrays have been very important in assessing the light transmission through well-defined geometries and dimensions, such apertures in flat silicon membranes suffer a severe problem when used to obtain scanning near-field images of real biological membranes. Specifically, it is hard to align these apertures with respect to a surface and it is hard to probe recesses in membranes with apertures in a large flat Si wafer. Nonetheless, aperture arrays can be effectively used in a unique nonscanning near-field method that should be very important to a whole class of problems in membranes and surfaces. We have called this method near-field statistical analysis of submicron structures (NSASS).[16]

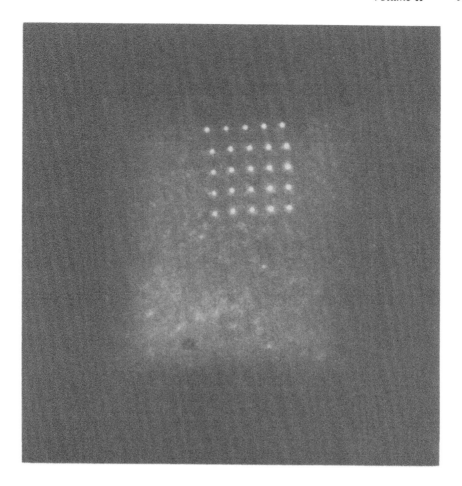

A

FIGURE 7.   Perylene fluorescence through (A) 2400-Å-, (B) 1200-Å-, and (C) 600-Å-diameter apertures are shown. The exciting laser light at 5145-Å illuminated the sample from below, and the fluorescence was detected through the apertures from above. The laser power that was used caused no detectable bleaching of the sample. The fluorescence was recorded with ASA 400 Kodak Ek-tachrome® film using a 30-sec exposure. These arrays contain graduated aperture sizes with the maximum corresponding to the diameters noted above.

This new method is based on the ability to assess the relative intensity of spectral phenomena, such as fluorescence, through each aperture in the array. Our arrays are ideal for this because each subwavelength aperture is separated by several microns. Hence the Airy discs from successive apertures do not overlap and, as can be seen in Figures 6 and 7, the intensity through each aperture can be readily assessed. A variety of systems could be studied with this technique. One specific example concerns model membrane systems which can be constructed as monolayers, bilayers, or multilayers with Langmuir-Blodgett techniques. These flat membranes can be placed directly on the aperture array or the array can be placed at some specified distance from the membrane using the micromovement technology discussed further on in this paper. Alternatively, preparations of membraneous organelles, such as chloroplasts, can be prepared and placed directly on the array or, within the near-field, as we will discuss in the next section.

From a statistical analysis of the fluorescence intensity transmitted through many apertures, both size and distribution information on fluorescent domains can be obtained. In Figure 8

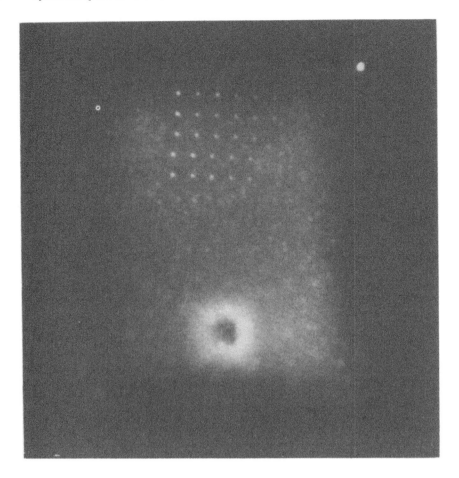

FIGURE 7B.

we show how one can graphically record this information. Specifically, the number of apertures is plotted versus the ratio of the fluorescence intensity (I) to the maximum fluorescence intensity obtainable (I_{max}). With such a graphical representation, let us consider two extreme limits of domain size that could be distinguished. In case A, where the domain size is much greater than the aperture dimension, a large proportion of apertures will have maximal fluorescence emanating from them. In case B, the domain size is much smaller than the aperture dimension, so that each aperture contains a similar number of domains. In such a case, the peak in the I/I_{max} axis occurs at a value related to the area fraction of fluorescence to total area. An important extension of such an experiment would be to use several arrays, each array having a different size for the individual apertures. The statistical analysis of the data from such experiments will allow us to obtain information on the distribution of domain sizes.

V. REACTION-CENTER SEGREGATION IN CHLOROPLASTS

We have already begun to test the NSASS concept with a chloroplast preparation applied directly to the apertures. Chloroplasts contain two photosystems, Photosystems I and II, which are thought to be segregated in the chloroplast membrane,[17,18] and NSASS can be used to test this idea under normal environmental conditions and without chloroplast disruption.

Photosystem II is involved with the splitting of water, while photosystem I is involved

FIGURE 7C.

with the alteration in the oxidation-reduction potential of electrons ejected from the split water. These electrons eventually reduce CO_2 to form glucose in the dark reactions of the Calvin cycle. Both these photosystems are found in the thylakoid membranes of chloroplasts. The thylakoid membranes contain two structurally distinct regions. These are called the grana, in which the membranes are stacked, and the stroma lamella, which is a region of unstacked membranes, interconnecting the grana regimes. A variety of experimental evidence indicates that there is extreme lateral heterogeneity in the distribution of Photosystems I and II.

NSASS can be readily applied to the problem of chloroplast segregation since, at room temperature, the fluorescence emanating from chloroplasts is exclusively from Photosystem II. In addition, the reagent 3-(3,4-dicholorophenyl)-1,1-dimethylurea (DCMU) can be used to enhance Photosystem II fluorescence by blocking electron transfer from Photosystem II to I. Therefore, viewing in near-field, through an aperture array, the fluorescence from Photosystem II should provide us with an important probe of this segregation phenomenon. We have in fact been able to selectively and nondestructively view Photosystem II fluorescence through a 2400-Å aperture array using exciting light at 5145 Å from an argon ion laser with an incident power of 0.067 $\mu W/\mu m^2$. The results of this experiment are seen in Figure 9. These results were obtained in epi-illumination. Such an illumination and detection scheme is preferable because the intensity of the radiation field in the near-field exponentially increases as the object is brought into near contact or contact with the aperture. Therefore,

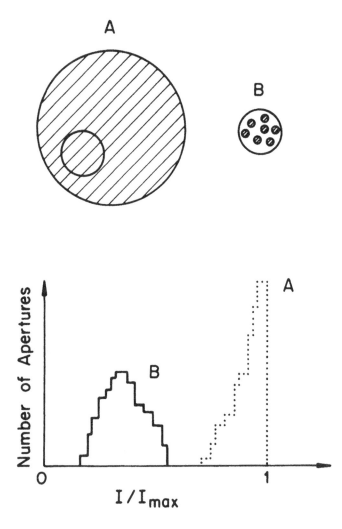

FIGURE 8.   NSASS. The schematic illustrates what we expect to detect in the two extreme cases of this nonscanning near-field imaging method. (A) The domain size is much greater than the aperture diameter. (B) The domain size is much smaller than the aperture diameter.

not only is there effective collimation of the incident radiation but there is also an effective intensity increase in the light beam.[6]

The data in Figure 9 clearly indicate heterogenous fluorescence intensities through the various apertures. However, before these results can be interpreted in terms of support for a segregation model of chloroplast structure, additional experiments have to be completed. Included among these experiments are the determination of relative chloroplast density at each aperture and a repetition of the above experiments under ionic conditions in which chloroplast membranes are thought to unstack and produce a homogeneous distribution of photosystems. These experiments are currently underway in our laboratory and the results should yield an unequivocal determination of photosystem segregation without chloroplast disruption and without the need for using probes that require nonphysiological environments.

VI. NEAR-FIELD SCANNING OPTICAL MICROSCOPY

The most general form of near-field imaging has been called near-field scanning optical

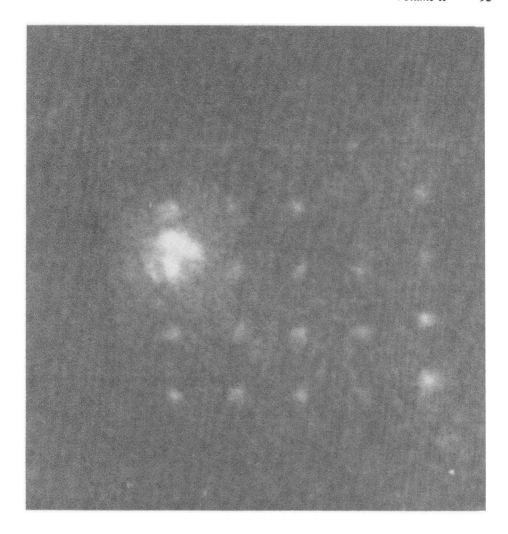

FIGURE 9. Photosystem II fluorescence is shown detected through 2400-Å-diameter apertures using epi-illumination. The incident intensity from an argon ion laser at 4579 Å was 0.067 μW/μm^2 at the surface of the array. Kodacolor® ASA 400 film and an exposure time of 40 sec was used to detect the fluorescence from the chloroplasts.

microscopy (NSOM).[4-6] To effectively apply NSOM to biological problems, two problems had to be addressed. The first of these problems is that real-cell membranes are rough. Thus, if one is to maintain the near-field condition in recessed pits and crevices of cell membranes, planar aperture arrays have to be replaced with a well-defined aperture at the tip of an extended probe. The second problem concerned the accurate determination of NSOM resolution before any applications to biological problems could be envisioned. For resolving the question of resolution, it was imperative to microfabricate well-defined patterns. Both these problems of test pattern and aperture fabrication have now been solved and our solutions will be discussed in the following sections.

A. Pipette Apertures

We have borrowed heavily on the biophysical technique of patch clamping[19] in order to overcome the requirement to construct apertures capable of probing nonplanar cell membranes. Specifically, we realized that patch-clamping pipettes appropriately modified could make ideal apertures at the tip of a long probe. Our approach has extended the patch-

Micropipette   Fabrication

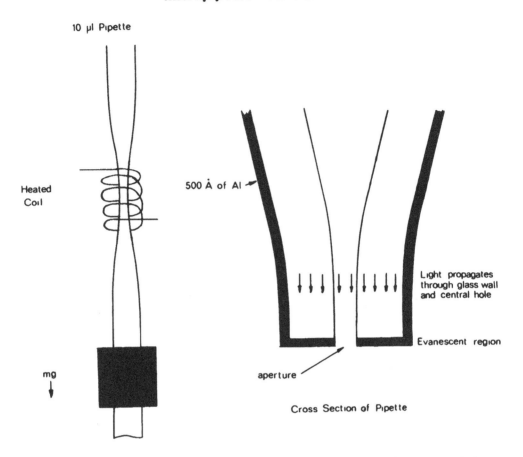

FIGURE 10.   A schematic representation of the procedure used for microfabricating the pipette apertures is shown on the left. On the right, a cross-sectional representation of the metallized pipette is shown together with the propagation of light toward the tip.

clamping pipette technology in four ways. First, we have been able to reproducibly fabricate apertures with diameters of <1000 Å using a pipette puller built in our laboratory. Second, we have shown that metallization of these pipettes can be accomplished without obstructing the apertures. Third, we have characterized the apertures at the tips of the pipettes with scanning electron microscopy. Fourth, we have demonstrated that light is efficiently transmitted through such pipette apertures.

For constructing these apertures, a simple gravity-driven pipette puller was used (see Figure 10). This instrument pulled the pipettes in two stages, first tapering the pipettes, then pulling until the breakage point was reached. Using this method and by varying the wall thickness of the glass tubing used to make the pipettes, we have been able to construct apertures as small as 500 Å. More recently we have begun to attack the problem of the outer diameter of the pipettes and have produced wall thicknesses of <1000 Å.

The overall shape of the pipette seen in the scanning electron micrograph reproduced in Figure 11 appears to be important for the light transmission. Our pipettes have a large diameter which rapidly constricts at the tip. Thus, as shown in Figure 10, the light exists in a propagating mode through most of the length of the pipette, which is above the cutoff frequency calculated by treating it as a classical wave guide. Only in the thin metallized region at the tip is a region of evanescence reached.

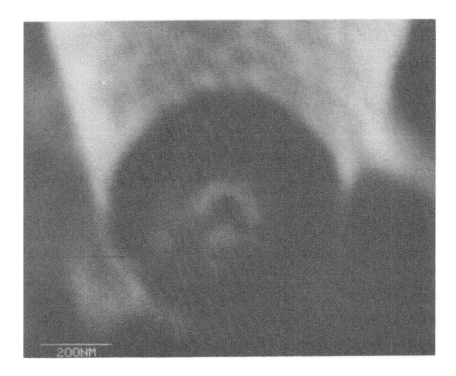

FIGURE 11. A scanning electron micrograph of the tip of a metallized pipette is shown with a 5000-Å outer diameter and with a <1000-Å central aperture. Although even smaller apertures could be prepared with our technique, they are difficult to characterize with the limited resolution of the SEM available.

B. Test Patterns

In order to accurately assess the resolution of any NSOM instrument it is essential to have microfabricated test patterns that can be well-characterized. To produce such test patterns, the same Si wafers and Si_3N_4 methodology used to prepare the apertures was employed. However, instead of aperture patterns, the scanning electron beam was used to inscribe gratings in the Si_3N_4 membranes. These gratings were then used as masks that were placed on glass slides and then aluminum was evaporated through the glass onto the glass slides in a contact printing procedure to create microfabricated aluminum gratings on glass. These gratings were characterized with scanning electron microscopy as shown in Figure 12 before any NSOM scans were attempted.

C. Prototype System

In order to test the concept of NSOM we have assembled a prototype system (see Figure 13). This system used a pipette-type aperture that was scanned with two piezoelectric crystals, one for z translation to bring the aperture into the near field and the other for translation in the x direction. The translation was controlled by an IBM PC computer which moved the aperture across the test grating in 300 Å steps. The light source illuminated the grating through the pipette and an objective lens imaged the whole grating onto the photocathode of an optical multichannel analyzer (OMA). Thus, as the pipette was translated across the grating the OMA spatially monitored the modulation of the transmission as the pipette was moved from the transparent to the opaque regions of the grating.

For these initial feasibility experiments the vibration isolation was a two-tiered system. The first stage was simply a large metal tabletop mounted on tennis balls. Measurements with a geophone demonstrated that such an isolation was sufficient to reduce vertical dis-

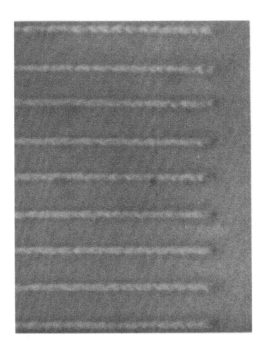

FIGURE 12.   A scanning electron micrograph of a
grating of aluminum lines on glass used for our resolution tests is shown. The lines were 2000 Å separated
by 4000 Å. To obtain this micrograph the sample was
coated with a 50-Å-thick film of gold palladium.

placements to less than 300 Å. A second-stage isolation system which was a steel plate on
a foam pad rested on the aforementioned tabletop. This second stage provided intermediate
frequency range isolation which helped isolate the microscope in a frequency range from
10 to 50 Hz. Although this level of isolation technology was adequate to demonstrate the
feasibility of the NSOM concept, it falls short of the current state of the art for a number
of reasons (e.g., no acoustic isolation, poor low-frequency isolation, poor vibration damping,
and no isolation from horizontal disturbances). A more sophisticated isolation system will
be needed to achieve the full potential of NSOM and this will be discussed in the following
pages.

D. NSOM Results

In spite of the nonideal conditions of NSOM experiments completed to date, we have
nonetheless been able to effectively test the concept in the optical regime. Our results clearly
indicate that with the inclusion of the above design considerations in a second-generation
instrument, our goal of better than 500-Å resolution should be readily achievable.

In this initial sequence of experiments we hoped to achieve a direct comparison of NSOM
scans with known aperture diameters and electron micrographs of our microfabricated test
patterns. This we have been able to achieve in recently reported experiments[6] which are
reproduced in Figure 14. For these experiments, a grating was used with 4000-Å-wide
aluminum lines spaced 2000 Å apart on glass. Scanning electron micrographs were obtained
of both the grating and the aperture. The micrograph of the grating is shown in Figure 12.
A micrograph of the pipette aperture similar to the one shown in Figure 11 was obtained.
The NSOM scan shown in Figure 14 was recorded with an aperture that was in contact with
the grating.

To obtain the scan reproduced in Figure 14 the aperture was brought into contact with

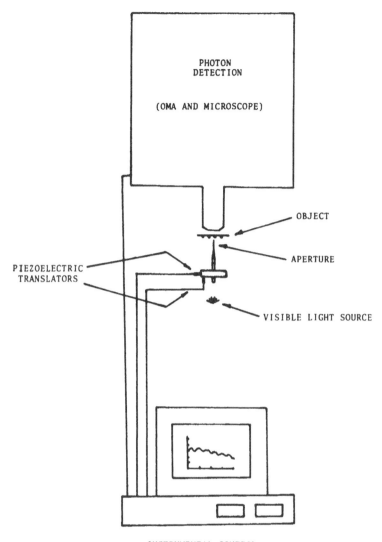

FIGURE 13. A schematic diagram of the experimental apparatus used for our initial feasibility experiments is shown. Both the optical multichannel analyzer and the piezoelectric translators are controlled with a microcomputer.

the grating under visual inspection. The aperture was then retracted from the grating using a calibrated piezoelectric crystal and then it was translated in the x-direction. Contact was once again made before a light-intensity reading was made. This procedure was followed during the whole course of the scan. A resolution of better than 1500 Å was achieved with a 1/2-μm-aperture diameter. This was deduced by averaging the sharpness of the steps in going from transparent to opaque regions of the grating. In spite of the size of the aperture, the resolution achieved is understandable in terms of the high signal-to-noise which permitted the detection of even partial aperture occlusion by the sharp edges of the opaque grating. The high signal-to-noise was indeed encouraging since the illumination source was not a laser for this experiment but rather light from a 100-W-tungsten lamp passed through the aperture of the pipette.

As mentioned above, an optical multichannel analyzer (OMA) with a silicon-intensified

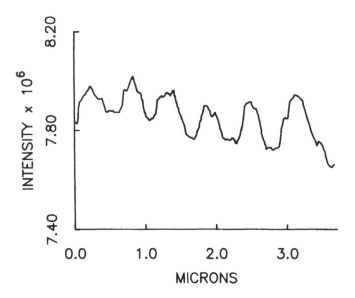

FIGURE 14. An optical scan of the test grating shown in Figure 12. From the sharpness of the intensity modulation between adjacent maxima and minima, a resolution of <1500 Å is inferred. Furtheremore, note that the periodicity of the grating is reproduced over the entire scan.

vidicon was used to detect the signal in these first experiments. The use of an OMA which can view the whole field of the objective lens of the microscope has some advantages over single-channel detection with a photomultiplier, but, these advantages are outweighed in NSOM by photomultiplier detection. In conventional fluorescence imaging, a two-dimensional digital recording device like an OMA is essential. However, in NSOM the aperture produces the digital image of the object as it scans across the surface, thus, there is no need to use an OMA. The photomultiplier has considerably lower dark count, has a wider dynamic range, and, unlike an OMA, has the capability to detect single photons. NSOM can use all these superior qualities of photomultipliers to generate high-resolution images of fluorescing objects.

VII. OTHER ATTEMPTS AT NEAR-FIELD IMAGING

Research into near-field imaging of fluorescence is currently centered at Cornell University, Ithaca, N.Y. In addition, until our 1983 report,[1] there were no published attempts in optical near-field imaging. Nonetheless, the idea of using for imaging the collimation of radiation in the near-field was first attempted in the microwave regime over a decade ago.[20] Ash and Nicholls[20] were able to use 3-cm microwaves to observe periodic oscillations as a result of scanning over an aluminum grating with separations and lines λ/60 in dimension. They were also able to form a two-dimensional image of the letters UCL with a resolution of λ/20. These pioneering experiments were not extended to the optical regime in the decade that elapsed between the Ash and Nicholls[20] and Lewis et al. reports.[1] This is understandable because of the dimensions involved in the optical-imaging experiments. It also should be noted that for microwave frequencies the boundary conditions that govern resolution are very different from those in the optical regime. Specifically, the completely conducting, thick, opaque, screens used for the apertures in the microwave regime are quite different from those employed with optical frequencies in which thin-metal films with finite conductivity are used.

Since our publication, there have been several additional reports using near-field collimation for imaging with optical[21,22] and far-infrared frequencies.[23,24] The experiments at optical frequencies were performed by our group,[2-6] by Pohl et al. at IBM Zurich,[21] and Fischer at the Max Planck Institute for Biophysical Chemistry at Gottingen.[22] Pohl et al.[21] have introduced a new method for constructing apertures which involves anisotropic etching of a single crystal of quartz with HF to form a pyramidal structure with a highly pointed tip. The quartz pyramid was then coated with metal and the tip was then thrust with a piezoelectric device called a bimorph against a glass slide while light was illuminating the unetched uncoated end of the pyramid. A sensitive photomultiplier detected when light began to propagate through the initially opaque tip at the apex of the pyramid as the tip was crushed against the glass. Although apertures produced by this procedure are not ideal, these workers felt that their results indicated a resolution of 250 Å. Fischer[22] scanned a subwavelength aperture over a larger aperture. However, the use of grazing incidence monochromatic radiation from a HeNe laser made the experiment particularly difficult to interpret, possibly because of the presence of a series of standing waves. Additional difficulties in this experiment included the use of thin-metal films that caused high-background transmission in the films with a lack of definition in the subwavelength aperture used. As noted above, it is important to stress accurate definition of apertures and gratings in terms of geometry, dimension, and efficiency of light transmission. We believe under these well-defined conditions <1500-Å NSOM resolution has clearly been shown with the clear potential of <500 Å resolution.

VIII. Z POSITIONING

A crucial step in achieving the vast potential of NSOM in biology is an independent accurate (± 30 Å) measurement of the aperture position in the z direction, a direction normal to the surface. The knowledge of aperture/surface separation is important for the incorporation of a feedback loop to ensure that the aperture is in the near-field. Previously, the near-field condition was met by simply achieving contact between the aperture and the object. For rough-cell membranes this is not easy to ensure, even with our tapered pipette-type apertures and it is a procedure that is destructive to both the aperture and the object.

To appreciate the stringency of the near-field condition we have considered,[6] theoretically, the transmission of light through a slit in an opaque screen. The theoretical techniques used were first formulated by Neerhoff and Mur.[25] For our application, a series of four coupled integral equations was formulated using Green's functions and then solved numerically to find the electromagnetic fields and energy flux in the near-field. In Figure 15 we plot the energy flux as a function of both the aperture plane and the z distance normal to the aperture plane. The aperture used in our calculations is an infinitely long 500-Å-slit in an 1800-Å-thick screen. The z distance considered in Figure 15 is 0 to 1000 Å, which is the region normally associated with the near-field for 5000 Å radiation. Two conclusions can be deduced for these results. First, within one slit radius, the radiation remains highly collimated to the dimension of the slit. Thus, for a 500-Å-slit the near-field extends $\simeq 250$ Å whereas for a slit of 2000 Å the near-field collimation is effective to $\simeq 1000$ Å. Second, in the region close to the slit, the presence of the slit causes a dramatic increase in the radiation-field intensity. This intensity increase exponentially decreases as a function of the direction perpendicular to the aperture plane. As can be seen from Figure 15, if intensity variations in the fluorescing object are to be meaningful as the aperture is scanned in the x-y plane it is imperative that stringent z control must be maintained.

How can a pipette-type aperture be made to follow the contour of the sample at a z distance constant to ± 30 Å? Some feedback mechanism is required that can rapidly adjust to changes in the aperture to cell-membrane separation. Such changes in separation can

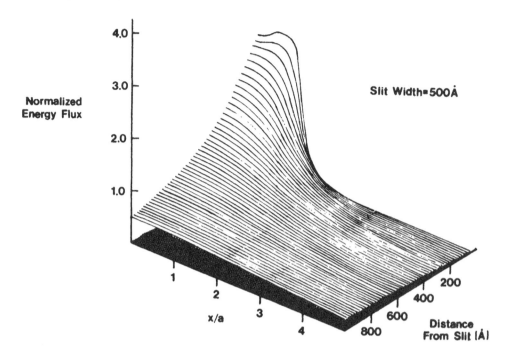

Normalized Energy Flux

Slit Width=500Å

x/a

Distance From Slit (Å)

FIGURE 15. A plot of the energy flux emanating from a 500-Å-diameter slit normalized to the energy of the plane wave incident on the slit. The normalized energy flux is plotted as a function of both the distance from the slit-containing screen and also across the slit (x/a = 0 is the slit center and x/a = 1 is the edge of the slit). Notice the collimation of the radiation and the exponential decrease in the near-field intensity.

occur for a variety of reasons, including contour variations as the aperture is moved across the surface or even rapid dynamic movements of the living-cell membrane as the aperture is recording at a single point. Not only must the feedback mechanism track such movements, but, it must accomplish this feat nondestructively, with an accuracy of ± 30 Å and at distances of 200 to 300 Å, which are characteristic of near-field collimation and efficient fluorophore to aperture energy transfer.

In fact, several methods exist for such feedback; these include: the measurement of capacitance changes between the aperture and the surface, the use of fluorescence, and the application of tunneling. In terms of capacitance, most cells have a surface negative charge and femto-farad (10^{-15} farad) capacitances have already been measured in cells.[12] When this is combined with the very recent development of a scanning capacitance microscope,[26] capacitance becomes a technique that fits many of the requirements for an effective feedback mechanism. Matey and Blanc, the inventors of scanning capacitance microscopy, have shown that if a capacitance probe is placed 200 Å from a surface, surface roughness of ± 2 Å can be detected, even though in this initial report of capacitance microscopy a high degree of lateral resolution was not achieved. It is likely that if our metallized pipettes are used we will be able to obtain significantly improved lateral-spatial resolution. Thus, we envision capacitance as allowing us to rapidly, within a time much less than the residence time of the aperture at each pixel, maintain our aperture at 200 Å from the surface with an accuracy of ± 30 Å. Capacitance measurements have the additional advantage of being nondestructive, since current densities are nominal and this has already been shown for the stimulation of mast cells.[12] In addition, capacitance should also provide an ultrahigh-resolution map of surface topography. It is even conceivable that we will be able to eventually obtain NSOM scans at different surface/aperture separations. This will allow optical sectioning of the near-field regime and could provide three-dimensional resolution within the near-field.

Although capacitance is a promising z-separation feedback mechanism, other techniques also exist which could maintain this separation to ± 30 Å. One such alternate method is to use to advantage the exponential change as a function of z of the near-field fluorescence intensity. To use the near-field fluorescence, the membrane needs to be labeled with two fluorophores with different emissions. One fluorophore is used for the lateral imaging, while a second fluorophore uniformly distributed among the membrane lipids is used for the z separation. Specifically, as the aperture approaches the surface the fluorescence should exponentially increase. Such a change in intensity of the uniformly distributed fluorophore can then be used to control the piezoelectric expansion of the pipette holder from 200 Å, thus forming a complete feedback system.

A third feedback mechanism is measuring the tunneling current from the surface. Scanning tunneling microscopy has achieved spectacular horizontal and vertical resolutions of 1 Å.[27] However, a primary disadvantage of tunneling lies in its extremely short effective range. In fact, simple calculations of tunneling through a work-function potential, which has been reduced by an applied field indicate that a current on the nanoamp level is obtained only when the pipette is roughly 20 Å away from the surface. This is a factor of ten closer than the separation which is needed for NSOM and the current densitives produced could be destructive to the membrane. In this regard it is worth noting that ion channels in membranes have conductances on the picoamp level which are 10^3 smaller than the current densitives which are presently used in tunneling.

IX. OTHER ESSENTIAL CONSIDERATIONS FOR SECOND-GENERATION NSOM

In addition to z positioning, a second-generation NSOM instrument must incorporate various important features that were not a part of our prototype system. These features, some of which are being incorporated in a second-generation system presently being completed in our laboratory, include various considerations on vibration isolation, aperture, and sample translation, signal detection, aperture fabrication, viewing modes, contrast enhancement, and other factors. Some of these considerations will now be discussed in detail.

A. Vibration Isolation

Minimal vibration isolation was used in our prototype system, but, to achieve the full potential of NSOM the vibrational isolation has to be improved. High-quality optical tables are available that can isolate from horizontal and vertical disturbances of 10 Hz and above. These tables provide excellent rigidity and good vibration-damping properties. However, in addition to such isolation, a rigid second-stage system must shield against the 5 to 10 Hz disturbances which would otherwise be transmitted by the optical table. The second stage must also isolate against vibrations not associated with the building. Such motion includes acoustic vibrations that directly cause disturbances in the second-stage platform on which the instrument is constructed. Other sources of these direct disturbances are relaxation of mechanical and thermal stress and motion of the scanning stage. Although damping for ultimate NSOM applicability may appear severe, it is important to realize that damping techniques have been devised for scanning tunneling microscopy that allow for stability at a 0.1-Å level for vertical disturbances and a 3-Å level for horizontal disturbances.[27]

B. Aperture and Sample Translation

To address the problem of micropositioning more fully, a sophisticated translation system is currently being developed for use in our second-generation instrument. Coarse positioning in the xy plane is again accomplished with two stacked precision stages. However, these stages are now driven by DC motorizers instead of the more conventional micrometers, which permit the operator to translate the sample under the pipette without inducing any unwanted motion in the second-stage platform. The use of motorizers is also advantageous because of their speed (greater than 100 μm/sec), long travel range (up to several inches), and accuracy (better than 1000 Å mechanical resolution).

For the fine positioning needed in the course of a scan, a recently developed,[28] commercially available, xy stage will be mounted atop the coarse stages. This stage has very small hysteresis (<1% of the full expansion), so that it is possible to scan in two dimensions with a high degree of reproducibility. Furthermore, the total range is greater than 50 μm in each of two directions and this exceeds that of more typical piezoelectrically driven stages. Since yaw, pitch, and roll motions are confined to less than one arcsecond, there should be negligible distortion and loss of contrast in the resulting NSOM images. The lack of noticeable lateral motions will also ensure that the relative aperture to object separation in the z direction is maintained to within the ± 30 Å mentioned previously. Finally, the resolution of the stage in the x and y directions is better than 10 Å. Hence, this combination of coarse and fine stages should satisfy any current or conceivable future needs for generating ultrahigh-resolution NSOM images.

The pipette in our second-generation instrument is held stationary in the xy plane above the sample, which is scanned while the z position of the pipette is adjusted. For the z positioning of the pipette, piezoelectric devices once again are used. Such devices have already been shown to have an accuracy in positioning of 0.1 Å.[27] Of particular concern in the design of the housing for these scanning stages is the minimization of thermal drift problems and we are confident that such problems are surmountable using appropriate shielding.

There are numerous advantages to the scanning system described here. First, the sample is moved rather than the pipette, so that no whip-like motion is induced in the rather flexible pipette tip. Second, if the sample is scanned, the optical path is stationary and the design of optical components for the detection of the signal in the far-field is greatly simplified. Third, in a scanning arrangement the field of view is only limited by the number of pixels available in the display system. Thus, even at super-resolution, the resolution of an NSOM system is independent of the field of view. Finally, since an electrical signal is detected point-by-point, it is possible to take advantage of a wide range of imaging mechanisms. In short, the NSOM method can make use of most of the advantages of scanning optical microscopy over conventional optical microscopy (see, for example, Wilson and Sheppard[29]) with the additional advantage that the resolution is virtually independent of the wavelength used.

C. Signal Detection

Although the design of our new signal-detection system will depend upon which of three possible viewing modes is employed (see discussion which follows), a sensitive photon-counting assembly will form the core of the system. As mentioned above in the Prototype System section, NSOM does not require a vidicon detection system but can use all of the advantages of a photomultiplier. One of the principal advantages of such high-sensitivity photomultipliers is their photocathode quantum efficiency, which is about 20%, or roughly three times more sensitive than that SIT vidicon used in the prototype instrument. Furthermore, by cooling the tube, the dark noise can be reduced to considerably less than ten counts/second. Hence, it will be possible to record the weak fluorescence from biological systems, while still limiting the incident illumination to nondestructive levels. Finally, the TTL output pulses from the amplifier/discriminator electronics attached to the PMT can be sent to the controlling microcomputer. Within this computer, the counts can be separated into bins on a millisecond time scale. Thus, with this detection system, it will be possible to generate images at the rate of hundreds of pixels per second, thereby circumventing the scan-speed limitations of our initial model.

D. Aperture Fabrication

Although the pipette-type apertures currently available should produce good-quality NSOM scans in a second-generation system, we are still attempting to optimize our aperture fabrication methods for near-field applications. For example, we have been experimenting with different pipette diameters, wall thicknesses, and pulling conditions in order to produce ever-smaller apertures.

Of equal importance to the question of aperture size is the issue of pipette tip diameter. As mentioned previously, one of the primary advantages of the pipette-type apertures is their ability to probe recessed regions of a nonplanar surface. However, even these apertures can scan over surfaces of only limited roughness, since, if the radius of curvature of a recessed valley on the surface of a sample is smaller than the radius of the pipette tip, it will be impossible to position the aperture region within the few hundred angstroms necessary for near-field imaging. Thus, the reduction of tip size is another goal of our continuing investigation of aperture fabrication processes.

For the electron beam-fabricated apertures, extremely small holes can already be formed, so that the emphasis for future research will be somewhat different. For example, we hope to use current microfabrication technology to produce highly tapered projections from the surface of a silicon nitride membrane. Apertures could then be made in these projections. In this manner, it would be possible to combine the main advantage of the pipette-type apertures (the ability to scan over modulated surfaces) with all the advantages of aperture arrays (the ability to form small highly reproducible apertures in a variety of shapes).

To further exploit the flexibility of the microfabrication methods available, we hope to produce apertures of carefully tailored shapes in order to increase the resulting throughputs. In particular, it may be possible to borrow from the impedance concepts of microwave antenna theory to increase the radiation transmitted by an aperture. For example, we could etch an undercut profile in a metal film so that the shape-resulting aperture resembles that of an exponential horn. In this case, the aperture may have the ability to funnel the incident radiation before sending it out in a highly collimated beam.

For both the pipette-type and electron beam-fabricated apertures, the nature of the evaporated metal film used to define the surrounding opaque screen is very important. Hence, we are using the known properties of these films (see Table 1) to design uniform apertures yielding high throughputs. From the standpoint of uniformity, we are searching for metals with the very small grain size needed to produce apertures with sharply bound edges and well-defined diameters. In particular, chromium has an opacity almost as good as that of aluminum, but with a much smaller grain size. Furthermore, grain size reduction might be accomplished by the *simultaneous* evaporation of two or more metals, thus producing a metallic alloy screen. Finally, by varying the speed and temperature at which the evaporations occur, it may be possible to find an optimal set of conditions leading to small grain size.

The *sequential* evaporation of two or more metals might be used to increase aperture throughputs. Again, this argument can be understood in terms of impedance concepts: if the incoming radiation is initially incident upon a region of low opacity, the small change in impedance should ensure that relatively little of the radiation is reflected away. As the transmitted radiation passes through layers of successively greater opacity, the aperture providing the collimation will become better defined, while the loss in intensity due to reflection (as opposed to absorption) should be reduced.

In short, the apertures we have already produced only scratch the surface of the available fabrication technology. In the future, this technology should produce apertures which yield both better signal-to-noise and higher resolution in NSOM applications.

E. Viewing Modes

In our initial NSOM prototype, a particular form of transmission microscopy was used in which light was radiated from a pipette tip to illuminate a small portion of the test grating within the near-field. The transmitted light was then collected on the other side of the sample with a high numerical aperture microscope objective in the far-field. However, there is reason to believe that such a geometry is not ideal, since all the light which is scattered within the large acceptance angle of the objective is collected in addition to the light which is transmitted without scattering. This scattered light is diffraction-limited and, hence, can be interpreted as a source of noise. When the pipette aperture is over a highly opaque region (e.g., the aluminum lines of the test object) this does not present a problem, since the radiation is backscattered away from the object. However, when the aperture covers a boundary between two regions with vastly different optical properties (e.g., at the edge of an aluminum line), relatively large amounts of light can be scattered. This effect would have a tendency to obscure the sharpness of edges. In other words, with this method of transmission microscopy, the apparent features observed at the edges of regions of sharp contrast may not correspond to the true features present at these edges. Indeed, this effect may be one of the reasons why ≈ 1500 Å resolution was observed in our initial experiments, whereas the lines of the test grating were produced with considerably better sharpness.

To avoid these problems, a different mode of transmission microscopy can be used in which light is collected by the pipette within the near-field. Conceptually, this mode of operation is more difficult to understand than the original one, in which the illuminated aperture could be viewed as a subwavelength-size light source. Nevertheless, the feasibility of the collection mode is understandable, since, by the Helmholtz principle of reciprocity,[30] the pattern for the collection of radiation by an aperture is the same as that for transmission by this same aperture. In other words, the extreme intensity of the radiation within the collimated near-field will ensure that the collected light will originate primarily from the small area directly below the aperture. Furthermore, because of the small aperture diameter, the pipette will have an extremely small acceptance angle obfuscating scattered light collection. Hence, in this new mode of transmission microscopy, truer images should be obtained, along with possibly higher resolution.

In many potential NSOM applications, for example, in the imaging of bone, it will be impossible to use transmission microscopy because of the thickness and/or opacity of the sample. For such cases, a reflection mode can be used. This method of operation faces two potential limitations. First, since the radiation must pass through the end of the pipette not once but twice, the signal will be relatively weak, as a result of double attenuation within the aperture. However, with the sensitive detection system described above, this should not prove to be troublesome. Second, in nonfluorescence applications when light is initially sent down the wide end of the pipette toward the aperture at the tapered end, some fraction of the light will be backscattered into the detector. Although the relative fraction of backscattered light may be small, it may completely swamp the signal, resulting from light reflected back up from the sample via the small aperture. One method of overcoming this problem is to modulate the sample position at a frequency that is much higher than the scan rate and detect the signal with phase-sensitive techniques. In fact, it was this method that was used by Ash and Nicholls[20] in the original microwave experiments. Alternatively, reflection fluorescence microscopy could be used. In this scheme, monochromatic light from a laser will be reflected off a dichroic mirror and sent down the pipette. The fluorescent light excited in the sample will travel back up the pipette and be transmitted through the dichroic mirror to the detector. Virtually all of the backscattered noise should be eliminated, since light at the laser frequency will be reflected off the mirror and away from the detector. As will be discussed in the following pages, the signal-to-noise expected from such an arrangement indicates the feasibility of the experiment.

There are also definite advantages to the use of reflection microscopy. For example, in the reflection mode, the aperture illuminates only a small section of the sample surface and collects light from this same small area. Hence, the near-field collimation phenomenon is employed in both directions. This should significantly reduce noise due to the spreading of radiation in thick samples (see discussion which follows), and may lead to a higher effective resolution. In addition, this is particularly advantageous in fluorescence NSOM since this mode of operation allows the interrogation with light of only 1 pixel at a time. This considerably reduces fluorophore bleaching problems.

F. Contrast Enhancement

A variety of contrast enhancement methods can be employed to increase the effective NSOM signal-to-noise ratio and resolution of the resulting images. These methods fall into the broad categories of digital image processing, enhancement through different illumination schemes, and fluorescence labeling, which is the focus of this paper.

Because NSOM images are recorded as a series of discrete elements (pixels), digital techniques can be used to increase contrast. For example, a grey-level display system can be used, in which case the slope at the midgrey level can be adjusted to produce pictures with optimal contrast. The sharpness of edges can also be increased by statistical differencing. In this method, the pixel values are weighted to the SD of the pixel values in some neighboring region. By using Fourier filtering, the high-frequency noise present in an image can be eliminated. These represent some of the simplest processing methods available. Actually, it should be possible to apply to NSOM all digital techniques which are currently utilized in scanning optical and electron microscopy.[29]

Various types of visual information can also be obtained in NSOM applications by using illumination schemes which have been developed for conventional microscopy. For example, by using polarized illumination or phase contrast methods, different information and possibly greater contrast can be obtained in some instances.[21]

Finally, as in the case of fluorescence, the sample itself can be prepared so as to increase the visual contrast.

X. SIGNAL-TO-NOISE CONSIDERATIONS

In order to consider whether it would be possible to apply NSOM to the imaging of clusters of membrane-bound fluorescent molecules, two extreme examples can be considered. The first is the case of a fluorescent lipid domain where all the lipid molecules are fluorescent. In this example, for a 500-Å aperture suspended over a completely fluorescent regime and considering the area covered by a lipid molecule to be ≈ 50 Å^2 there would be $\approx 10^3$ fluorescent lipids within the aperture. The other extreme is the problem of a large receptor molecule where only one molecule is within the aperture. A good example of such a molecule is the low-density lipoprotein (LDL) receptor. This is a large molecule of ≈ 250 Å in diameter and it can be loaded with approximately 30 fluorescent lipid molecules. In certain types of cells and under the appropriate conditions these receptors have a distribution in the membrane that allows single fluorescent LDL molecules to be detected. Therefore, they provide us with the data we need for calculating the signal-to-noise in such a situation with only 30 fluorescent lipids within the aperture.

A. Signal

From studies with such fluorescent LDL receptors[31] (cf. Chapter 10 in this Volume) it is known that *without* resorting to reagents, such as *n*-propyl gallate[32] or submitochondrial particles[33] to decrease photodestruction, an LDL receptor can be illuminated and the fluorescence readily detected for 1 min with 1 μW/μm^2 of incident laser power. From this point the calculation of the signal in a mode where the aperture collects the fluorescence from a membrane illuminated in transmission proceeds in four steps:

1.   A determination of how many incident photons are present
2.   How many photons are absorbed
3.   The photons that are emitted as fluorescence
4.   The transmission through the aperture

Step 1. Number of Incident Photons per cm² (N)
From above

$$1 \ \mu W/\mu m^2 = 1 \ \mu J/sec \ \mu m^2 \text{ intensity can be used}$$

Assume 2 eV/photon

$$\text{... \# of photons/sec } \mu m^2 =$$

$$10^{-6}(J/sec \ \mu m^2) \cdot 1/1.6 \times 10^{-19}(eV/J) \cdot 1/2(photon/eV)$$

$$\simeq 3 \times 10^{12} \text{ photons/sec } \mu m^2$$

Introducing a conversion to cm²

$$N = \# \text{ photons/sec cm}^2 \simeq 3 \times 10^{20} \text{ photons/sec cm}^2$$

Step 2. Number of Photons Absorbed

Given a cross section for absorption $= \sigma_A = 2.8 \times 10^{-21} \cdot (\epsilon_\lambda)$ cm²

where ϵ_λ has a magnitude of 10^4

$$\sigma_A = 3.8 \times 10^{-17} \text{ cm}^2$$

$$N' = \# \text{ of photons absorbed/sec} = \sigma_A \cdot N = 12 \times 10^3$$

Step 3. Introduction of Emission Quantum Yield (ϕ_f)
Step. 4. Transmission through Aperture (T)

$$\# \text{ photons/sec at detector} = N' \ \phi_f \ T = 12 \times 10^3(10^{-1})10^{-1} = 120$$

$$\# \text{ of photons/min} = 7200$$

B. Noise

In order to understand the sources of noise, other than the statistical photon-counting noise, which is $\sqrt{7200}$, a single LDL molecule is assumed to be within the aperture. In addition, except for internalized fluorescent receptors it is assumed that the fluorescence is restricted to the membrane and the selective illumination of the upper or lower surface of the cell can be achieved. Furthermore, recall that, as shown in Figure 15, the near-field intensity decreases exponentially. This exponential decrease is approximately proportional to $e^{-4\pi z/a}$ where "a" is the aperture diameter and z is the distance of the aperture from the surface. This ensures that only molecules which are very close to the aperture should significantly contribute to the fluorescence intensity. In addition, molecules restricted to the membrane in close proximity to the aperture, but not under the aperture opening, will have their fluorescence quenched by the metal. This metal-quenching of membrane fluorescence has been shown in numerous studies by H. Kuhn and co-workers.[9,10] Given these important characteristics of near-field fluorescence we calculate that, even without considering signal reductions due to solid angle or the $1/R^6$ dependence of the fluorescence emission, less than a 15% contribution to the fluorescence intensity can be expected from chromophores assumed to completely fill the cytoplasm and the "far-side" plasma membrane under the aperture. Assuming this worst-case situation and assuming a photomultiplier like a cooled RCA® C31034 with a dark count of 3/sec we calculate a signal-to-noise of $\simeq$6.

The above signal-to-noise was determined for a transmitted light geometry rather than epi-illumination. In this geometry the whole area is illuminated in transmission and the aperture is placed above the sample to detect the fluorescence from the opposite surface. The situation in epi-illumination should cause significant improvements. First, since only that portion of the membrane directly below the aperture is exposed at any one time, a much higher incident intensity can be used before the bleaching threshold is reached. Second, the exponential decrease in intensity within the near-field ensures that only the fluorophores directly below the aperture are strongly illuminated. Third, there is a decrease in noise because the aperture acts as a collector which passes photons preferentially from fluorophores in close proximity to the aperture. This is a result of the same exponential effect. Furthermore, if we consider the real situation of nonuniform distribution of fluorophores in the cytoplasm, the $1/R^6$ fluorescence-intensity falloff, and, if we consider the acceptance-angle effect, the signal-to-noise ratio increases fourfold over the above transmission case. Therefore, the time spent at each pixel can be reduced by approximately a factor of 16 to keep the same signal-to-noise of 6. With such a reduction in signal to noise a 100 pixel $\times$ 100 pixel image can be generated in $\simeq$25 sec.

XI. SUMMARY

In summary, fluorescence NSOM when coupled to a rapid, vertical adjustment feedback mechanism is going to be an extremely powerful, high-resolution, nondestructive probe of cells and cellular organelles. Specifically, the signal from the feedback mechanism can be used to obtain a super-resolution topographical map of the cell surface, while a fluorescence image through the scanning aperture is simultaneously generated. In addition, this technology which is being developed to realize the full potentials of NSOM can also be readily extended to test other suggestions, such as the detection of vibrational spectra in a 5-nm region using surface-enhanced linear and nonlinear Raman spectroscopy.[34] By depositing a silver particle at the tip of the pipette used in the NSOM experiment or by using the pipette tip directly, such Raman spectra in localized regions could be enhanced. In addition, by scanning the Raman probe it may be possible to produce a high-resolution map of the vibrational properties of the membrane. Thus, we are at the crossroads of a series of crucial developments that should revolutionize microscopy and spectroscopy. All this is possible because of the revolution in microfabrication, micromovement, low-light level detectors, lasers, and the availability of powerful microcomputers. There is no doubt that a combination of these highly compatible developments in scanning spectroscopies and microscopies will bring us to a new threshold in super-resolution, nondestructive, spectral, and kinetic imaging of biologically important macromolecular assemblies in native aqueous environments.

NOTE ADDED IN PROOF

Since this article was written there have been many important developments. By way of example the reader is referred to papers by A. Harootunian et al.[35] and E. Betzig et al.[36]

ACKNOWLEDGMENTS

The chloroplast experiments are being performed in collaboration with Prof. Chanoch

Carmeli who was visiting our laboratory from Tel Aviv University. The authors are grateful for support from the National Science Foundation Grant #ECS 84-10304 and the U.S. Air Force Contract #AFOSR 84-0314. Support of the National Science Foundation for the National Research and Resource Facility for Submicron Structures through Grant #ECS 8200312 is also gratefully acknowledged.

REFERENCES

1. **Lewis, A., Isaacson, M., Muray, A., and Harootunian, A.,** *Biophys. J.,* 41, 405a, 1983.
2. **Lewis, A., Isaacson, M., Harootunian, A., and Muray, A.,** *Ultramicroscopy,* 13, 227, 1984.
3. **Harootunian, A., Betzig, E., Muray, A., Lewis, A., and Isaacson, M.,** *J. Opt. Soc. of Am.,* A1, 1293, 1984.
4. **Betzig, E., Harootunian, A., Lewis, A., and Isaacson, M.,** *Biophys. J.,* 47, 407a, 1985.
5. **Betzig, E., Harootunian, A., Kratschmer, E., Lewis, A., and Isaacson, M.,** *Bull. Am. Phys. Soc.,* 30, 483, 1985.
6. **Betzig, E., Lewis, A., Harootunian, A., Isaacson, M., and Kratschmer, E.,** *Biophys. J.,* 49, 269, 1986.
7. **Betzig, E., Harootunian, A., Lewis, A., and Isaacson, M.,** *Appl. Opt.,* 25, 1890, 1986.
8. **Kuhn, H.,** *J. Chem. Phys.,* 53, 101, 1970.
9. **Kuhn, H.,** Spectroscopy of Monolayer Assemblies Part I, Principles and Applications, in *Physical Methods of Chemistry,* Part III B, Weissberger, A. and Rositer, B. W., Eds., Wiley-Interscience, New York, 1972, 579.
10. Drexhage, Interaction of light with monomolecular dye lasers, in *Progress in Optics,* Vol. 12, Wolf, E., Ed., North-Holland, Amsterdam, 1974, 163.
11. **Kunz, R. E. and Lukosz, W.,** *Phys. Rev. B.,* 21, 4814, 1980.
12. **Fernandez, J. M., Neher, E., and Gomperts, B. D.,** *Nature (London),* 312, 453, 1984.
13. **Finne, R. and Klein, D.,** *J. Electrochem. Soc.,* 114, p. 965.
14. **Isaacson, M. and Muray, A.,** *J. Vac. Sci. Technol.,* 19, 1117, 1981.
15. **Adrejewski, W.,** *Z. Angew. Phys.,* 5, 178, 1953.
16. **Betzig, E., Lewis, A., Harootunian, A., and Isaacson, M.,** to be published.
17. **Anderson, J. M.,** *FEBS Lett.,* 124, 1, 1981.
18. **Barber, J.,** *FEBS Lett.,* 118, 1, 1980.
19. **Sakmann, B. and Neher, E.,** Eds., *Single-Channel Recording,* Plenum Press, New York, 1983.
20. **Ash, E. A. and Nicholls, G.,** *Nature (London),* 237, 510, 1972.
21. **Pohl, D. W., Denk, W., and Lanz, M.,** *Appl. Phys. Lett.,* 44, 652, 1984.
22. **Fischer, U. Ch.,** *J. Vac. Sci. Technol.,* B3, 386, 1985.
23. **Massey, G. A.,** *Appl. Opt.,* 23, 658, 1984.
24. **Massey, G. A., Davis, J. A., Katnik, S. M., and Omon, E.,** *J. Opt. Soc. Am.,* A1, 1259, 1984.
25. **Neerhoff, F. L. and Mur, G.,** *Appl. Sci. Res.,* 28, 73, 1973.
26. **Matey, J. R. and Blanc, J.,** *J. Appl. Phys.,* 57, 1437, 1985.
27. **Binnig, G. and Rohner, H.,** *Helv. Phys. Acta,* 55, 726, 1982.
28. **Scire, F. E. and Teague, E. C.,** *Rev. Sci. Instrum.,* 49, 1735, 1978.
29. **Wilson, T. and Sheppard, C.,** *Theory and Practice of Scanning Optical Microscopy,* Academic Press, New York, 1984.
30. **Born, M. and Wolf, E.,** *Principles of Optics,* 4th ed., Pergamon Press, Oxford, 1970, 381.
31. **Barak, L. S. and Webb, W. W.,** *J. Cell Biol.,* 90, 595, 1981.
32. **Sedat, J.,** *Science,* 217, 1252, 1982.
33. **Bloom, J. A. and Webb, W. W.,** *J. Histochem. Cytochem.,* 32, 608, 1984.
34. **Wessel, J.,** *J. Opt. Soc. Am.,* B2, 1538, 1985.
35. **Harootunian, A., Betzig, E., Isaacson, M., and Lewis, A.,** *Appl. Phys. Lett.,* 49, 674, 1986.
36. **Betzig, E., Isaacson, M., and Lewis, A.,** Collection mode near-field scanning optical microscopy, *Appl. Phys. Lett.,* in press, 1987.

Chapter 13

FLUORESCENT PROBES OF ANION TRANSPORT INHIBITION IN THE HUMAN RED CELL MEMBRANE

James A. Dix

TABLE OF CONTENTS

I. INTRODUCTION

The human red cell has long been a popular system for scientific study because of its availability and simplicity, as well as its physiological importance as an integral part of the circulatory system. Among the myriad of red cell properties that have been studied have been oxygen binding to hemoglobin, water and nonelectrolyte transport, and cation and anion transport. In this chapter, we will focus on the anion transport function of red cells; in particular, we will describe applications of fluorescence spectroscopy in determining properties of inhibitor binding to the anion transport protein.

A. Physiological Importance

Anion transport in human red cells plays an important role in the efficient removal of carbon dioxide in tissue and the release of carbon dioxide in lungs. Carbon dioxide produced by metabolic processes diffuses rapidly into the interior of red cells, where it is hydrated to carbonic acid by the enzyme carbonic anhydrase. The resulting bicarbonate ion (the released proton binds to hemoglobin, thereby enhancing the release of bound oxygen) is transported out of the cell in a one-for-one exchange with extracellular chloride. In the lungs, the process is reversed: bicarbonate exchanges into the cell and is converted into carbon dioxide, which then diffuses out of the cell and lung capillaries, and into the atmosphere. This process is responsible for up to 80% of carbon dioxide transport in humans.

B. Mechanism of Anion Transport

There is good evidence[1,2] that anion transport occurs by a ping-pong mechanism (Figure 1; for enzyme terminology, see Segel[9]). This model envisions a transport protein that can exist in two conformations, one with the anion transport site facing the outside of the cell (E_o) and one with the site facing the inside of the cell (E_i). The transition between E_o and E_i can occur only if an anion is bound, thus accounting for the high value of 10^4 for the ratio of anion exchange flux to net anion conductance.[1] The transport sites can be "recruited" from one side of the membrane to the other by appropriate anion gradients.[4,5]

There is widespread agreement that red cell anion transport is mediated by a major transmembrane glycoprotein, Band 3.[1,2] Band 3 has a molecular weight of 95,000 daltons and is composed of a water-soluble cytoplasmic domain (43,000 daltons) and an integral membrane domain (52,000 daltons); the anion transport function is contained within the integral membrane domain. Although Band 3 exists in the red cell membrane as a dimer or higher multimer,[6] each monomer of Band 3 appears to transport anions independently of its neighbors.[7,8] A wide variety of anions are transported by Band 3, including monovalent and divalent inorganic and organic anions, with transport rates for the different anions that vary over four orders of magnitude.

C. Anion Transport Inhibition

Although numerous compounds have been shown to inhibit red cell anion transport, the disulfonic stilbene class of inhibitors (Figure 2) has proven to be the most useful in the study of transport. Certain disulfonic stilbenes bind almost exclusively to Band 3 and appear to inhibit transport competitively by binding to the transport site only when the transport protein is in the outward facing conformation (E_o in Figure 1).

Of the disulfonic stilbenes listed in Figure 2, the most directly useful for fluorescence studies of reversible transport inhibition have been 4-benzamido-4'-amino-2,2'-disulfonic stilbene (BADS) and 4,4'-dibenzamido-2,2'-disulfonic stilbene (DBDS). These compounds, like many of the disulfonic stilbenes, are weakly fluorescent in water, but unlike the other stilbenes, they undergo an increase in quantum yield of approximately 70 (BADS) and 80 (DBDS) when bound to Band 3 at 25°C. Thus, the fluorescence of BADS and DBDS

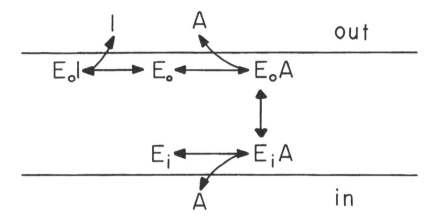

FIGURE 1. Mechanism of anion transport in human red cells. The figure represents a ping-pong mechanism in which the transport protein can exist with the anion binding site facing the extracellular solution (E_o) or the intracellular solution (E_i). The transition between E_o and E_i is allowed only if an anion (A) is bound, thus accounting for the one-for-one exchange of anions. From inhibition studies, disulfonic stilbenes (I) are thought to inhibit by binding to the extracellular transport site.

provides a convenient measure of the amount of inhibitor bound to the transport protein in equilibrium-binding experiments, and the time course of fluorescence increase provides a convenient measure for the kinetics of binding. Equilibrium binding and kinetics of binding can still be studied for disulfonic stilbenes that do not undergo fluorescent enhancement by monitoring intrinsic tryptophan fluorescence; Band 3 contains approximately 13 tryptophan residues, some of whose fluorescence is sensitive to stilbene binding.

In addition to studies with reversible transport inhibitors, it has also been useful to study irreversible inhibitors, in which the inhibitor reacts irreversibly at or near the transport site. (The term "covalent" is often used to denote irreversibility; this terminology is imprecise since covalent bonds can be reversible.) Irreversible inhibitors of red cell anion transport usually contain the isothiocyano group, which presumably reacts with lysine amino groups of the membrane protein. Irreversible inhibitors have the advantage that they remain in place throughout biochemical manipulations; by using a radioactively labeled irreversible inhibitor, such as tritiated 4,4'-diisothiocyanio-2,2'-disulfonic dihydrostilbene (H_2DIDS), the binding site can be localized to a particular protein, and, by proteolysis, to a particular protein fragment. The fluorescent irreversible inhibitor, 4-benzamido-4'-isothiocyano-2,2'-disulfonic stilbene (BIDS), has been used extensively for localization and energy transfer studies of Band 3.

In the following four sections, we describe the results of fluorescence studies designed to elucidate the environment, localization, and mechanism of inhibitor binding, and applications of fluorescent anion transport inhibitors to probe the structure and function of Band 3.

II. INHIBITOR ENVIRONMENT

The synthesis of the disulfonic stilbenes, BADS and DBDS, was described by Kotaki et al.[9] in 1971; these workers found a considerable fluorescence enhancement when BADS and DBDS were bound to human serum albumin. Cabantchik and Rothstein[10] reported in 1972 that BADS inhibited anion transport in red cells and fluoresced with increased quantum yield in the presence of isolated red cell membranes. A comparative study of the inhibition of anion transport by ten different disulfonic stilbenes was done by Barzilay et al.[11] who

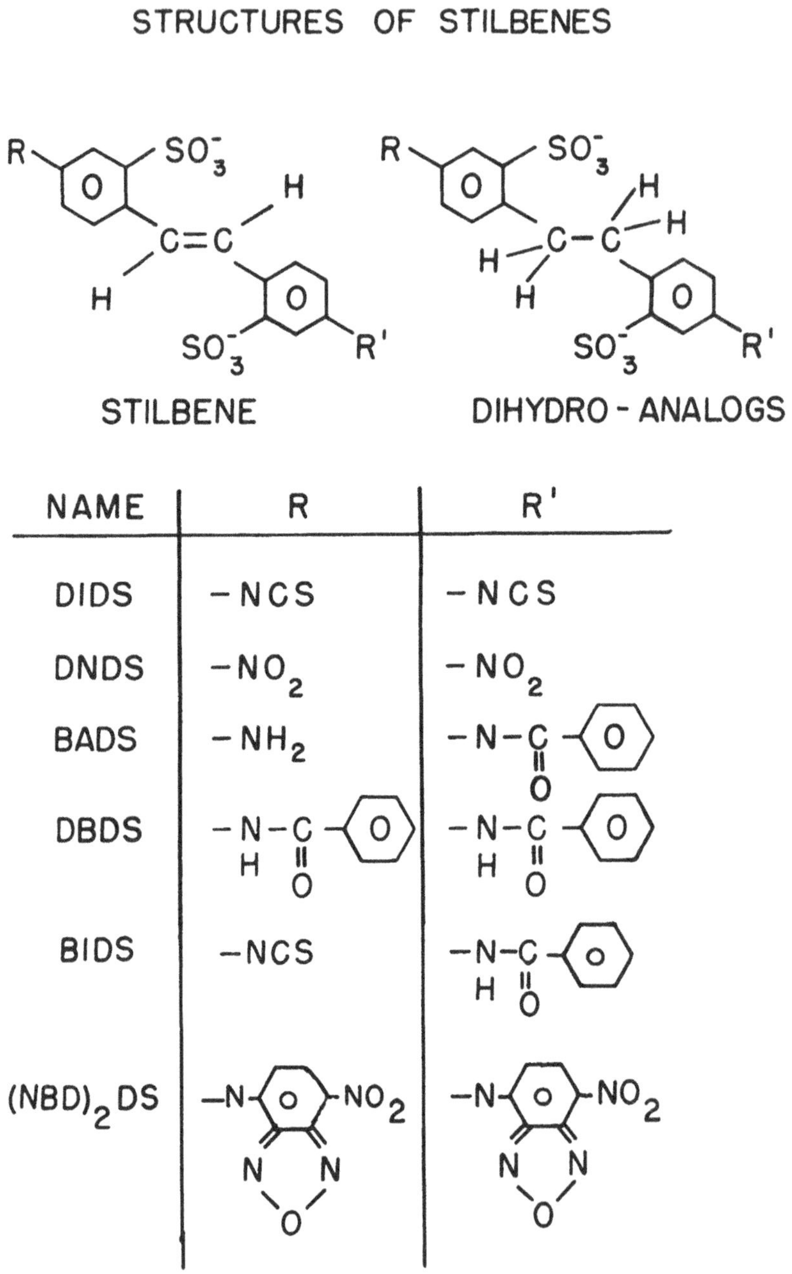

STRUCTURES OF STILBENES

STILBENE   DIHYDRO - ANALOGS

NAME	R	R'
DIDS	— N C S	— N C S
DNDS	— N O$_2$	— N O$_2$
BADS	— N H$_2$	— N — C ⬡(O) (C=O)
DBDS	— N — C ⬡(O) (H, C=O)	— N — C ⬡(O) (H, C=O)
BIDS	— N C S	— N — C ⬡(o) (H, C=O)
(NBD)$_2$DS	— N ⬡ NO$_2$ (N–N, O)	— N ⬡ NO$_2$ (N–N, O)

FIGURE 2.   Structures of some disulfonic stilbenes and their dihydro analogs.

correlated the ID$_{50}$ with the Hansch and Hammet parameters.[12] The results of this analysis indicated the disulfonic stilbene binding site on Band 3 is a hydrophobic pocket having electron donor capacity.

The fluorescence parameters of BADS and DBDS were exploited by Cantley and co-workers[13] to define in more detail the environment of bound BADS and DBDS. BADS exhibits a blue shift in its fluorescence-emission maximum when bound to the red cell membrane (Table 1). This blue shift is consistent with the conclusion, drawn from the Hansch and Hammet parameters, that the probes reside in a hydrophobic environment,

Table 1
PROPERTIES OF SOME FLUORESCENT DISULFONIC STILBENES

Compound	Solution				Bound to band 3						
	λ_{abs}^{max} (nm)	ϵ^{max} (M^{-1} cm^{-1})	λ_{emiss}^{max} (nm)	ϕ^a	λ_{abs}^{max} (nm)	ϵ^{max} (M^{-1} cm^{-1})	λ_{emiss}^{max} (nm)	ϕ^a	τ^b (nsec)	$(r_n^\infty)^c$	$(K_d)^d$ (μM)
BADS	340	34,000	481	0.0005	342[e]	—	465	—	—	>0.35	1.3
DBDS	336	50,000	416	—	336[e]	—	435	—	0.2	>0.35	0.068 0.82
BIDS	339	62,000	—	—	334[f]	51,000	435[f]	0.16	0.81	0.345	—

[a] Quantum yield.

[b] Fluorescence lifetime.

[c] Steady-state anisotropy extrapolated to immobilized Band 3 by a Perrin plot, as described by Dale et al.[20] The limiting anisotropy was equal to that of immobilized fluorophore.

[d] Equilibrium dissociation constant in 28.5 mM sodium citrate, pH 7.4, 20°C (BADS) or 25°C (DBDS).

[e] Total membrane protein concentration = 0.15 mg/mℓ, 28.5 mM sodium citrate, pH 7.4.

[f] Measured as a lysine adduct.

although the contribution of specific solvent effects and nonadiabatic relaxation processes[14] to the shift have not been assessed. The origin of the unusual red shift of DBDS when bound is not known. The quantum yields of BADS and DBDS increases upon binding to Band 3, and the anisotropy in the limit of immobilized Band 3 is equal to that in the limit of immobilized probe. These findings are consistent with the probe being rigidly held in its binding site.

III. LOCALIZATION OF INHIBITOR BINDING SITE

A. Energy Transfer Measurements

Band 3 consists of two easily separable and apparently functionally independent domains, a 52 kdaltons domain that resides in the membrane and is responsible for anion transport, and a 43 kdaltons domain which is thought to be exposed to or reside in the cytoplasm.[15,16] The integral membrane portion of Band 3 presumably contains the lysine groups reactive to the isothiocyano moiety, since labeling studies have shown it contains the binding site for tritiated H_2DIDS.[17] The cytoplasmic domain contains five sulfhydryl groups that are reactive to N-ethylmaleimide (NEM) in intact red cells (which are reduced to three in isolated red-cell ghost membranes, probably due to disulfide bond formation).[18,19]

Rao et al.[13] have used the different reactive sites in the two domains of Band 3 to measure, by fluorescence energy transfer methods, the distance between a sulfhydryl site in the cytoplasmic domain and the disulfonic stilbene binding site in the integral membrane domain. This study was particularly thorough in that three different methods of measuring the energy transfer efficiency were used: donor-quenching, acceptor enhancement, and donor lifetime. Furthermore, by suitable choice of maleimide reagent, Rao et al. were able to measure the energy transfer both from the cytoplasmic domain to the integral membrane domain and from the integral membrane domain to the cytoplasmic domain. The stilbene site was probed with 4,4'-dinitro-2,2' disulfonic stilbene (DNDS), DBDS, BADS, and BIDS; the sulfhydryl site was probed with 4-nitrobenz-2-oxa-1,3-diazole (NBD) and N-fluorescein-5-maleimide (NFM) (for transfer from stilbene to maleimide) and with N-[p-(2-benzoxazolyl)phenyl] maleimide (NBPM) and N-prenemaleinade (NPM) (for transfer from maleimide to stilbene). Structures and spectral properties of some sulfhydryl reagents are given in Figure 3.

There were several possible complications from the study of Rao et al. The maximum incorporation of the fluorescent maleimides was approximately two per Band 3 monomer. This finding, coupled with the fact that Band 3 is a dimer or higher oligomer in the membrane,[6] complicates the distance determination since the position of the cytoplasmic "anchor" is uncertain and energy transfer among monomers of the oligomer could occur. These problems were addressed by considering energy transfer with multiple donors and acceptors; the resulting uncertainty in the calculated distance amounted to 10 to 20%.

A second major problem is the determination of the orientation of the excitation and emission dipoles of the donor and acceptor, respectively, which is reflected in the value of κ^2. Dale et al.[20] has shown that local rotation of donor and acceptor molecules considerably reduces the error associated with an unknown κ^2. Although the limiting anisotropies of the bound stilbenes are large (Table 1), indicating restricted rotation, the limiting anisotropies of the bound maleimides ranged from 0.143 to 0.22 and were significantly less than those of completely immobilized maleimide. The rotational freedom of the maleimide reagents allowed a narrower range of κ^2 values.

In spite of these difficulties, Rao et al. found that the distance from the maleimide binding site to the stilbene binding site was remarkably independent of the method of measurement and of the nature of the donor-acceptor pair. Based on 14 different experimental conditions, the distance was computed to be in the range 3.4 to 4.2 nm. This distance is relatively short. Since the stilbene site is accessible from only the extracellular surface and the maleimide

NAME	STRUCTURE	λ_{abs}^{max} (nm)	ϵ (mM^{-1}cm^{-1})	λ_{em}(nm)
NBD-CL	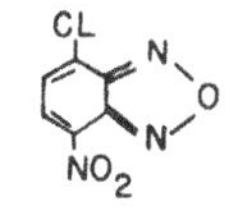	420	13	540
NBPM	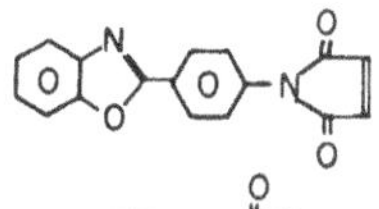	308	32	370
NPM	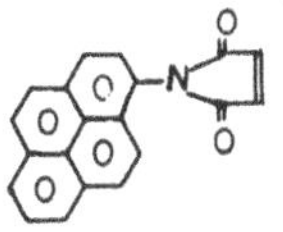	343 325 313	38	380 400 420
NFM	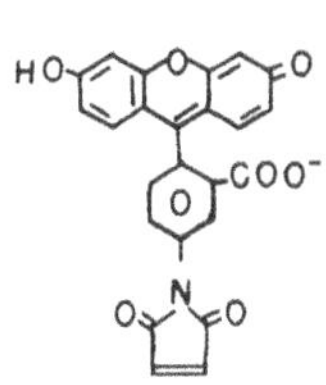	488		520
E5M	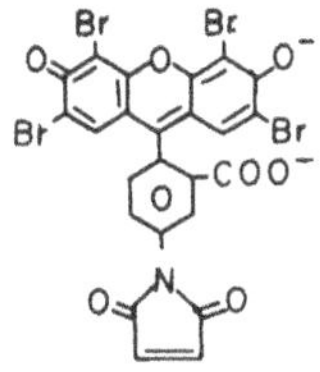	522	83	550

FIGURE 3.   Structure and optical properties of fluorescent sulfhydryl reagents known to bind to the anion transport protein, Band 3. Abbreviations: NBD-Cl, 7-chloro-4-nitrobenz-2-oxa-1,3-diazole; NBPM, *N*-[*p*-(2-benzoxazoyl)phenyl] maleimide; NPM, *N*-pyrene maleimide; NFM, *N*-fluorescein-5-maleimide; E5M, eosin 5-maleimide. (Reprinted with permission from Rao, A., Martin, P., Reithmeier, R. A. F., and Cantley, L. C., *Biochemistry*, 18, 4505, 1979, and Macara, I. G. and Cantley, L. C., *Biochemistry*, 20, 5096, 1981. Copyright 1979 and 1981 American Chemical Society.)

site is on or in the cytoplasmic side, these results suggest that the stilbene binding site is located some distance in from the extracellular surface.

B. Subunit Interactions

There is good evidence that Band 3 is a dimer or higher multimer in the red-cell membrane.[6] Interactions between sites on separate monomers have been observed with several different fluorescence methods.

Dissing et al.[21] used the probe, fluorescein mercuric acetate (FMA), to detect Band 3 interactions in red-cell membranes that had been stripped of other proteins. The mercuric moiety reacts with protein sulfhydryl groups to form mercaptide bonds. The labeling stoichiometry for FMA binding was consistent with a single site per Band 3 monomer. As seen in Figure 4, there is considerable overlap between the absorption spectrum and fluorescence emission spectrum of FMA, so that resonance self-energy transfer could be observed. The energy transfer was detected by depolarization of the acceptor. The data indicated that cytoplasmic domains of two Band 3 monomers are within 5 nm of one another.

That the disulfonic stilbene binding sites on two different Band 3 monomers can interact has been shown by several different fluorescence studies. Dix et al.[22] and Verkman et al.[23] studied the binding of the reversible transport inhibitor, DBDS, to Band 3 in isolated ghost membranes. Binding was monitored by fluorescence enhancement with corrections for flu-

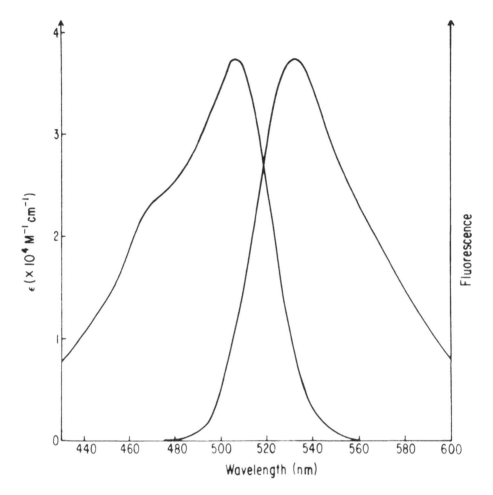

FIGURE 4.   Absorption spectrum and emission spectrum of fluorescein mercuric acetate (FMA) bound to Band 3. (From Dissing, S., Jesaitis, A. J., and Fortes, P. A. G., *Biochim. Biophys. Acta,* 553, 66, 1979. With permission.)

orescence from unbound DBDS, and inner filter and self-quenching effects. Control experiments showed that the fluorescence enhancement was completely abolished by pretreatment with the irreversible inhibitor, DIDS. Significant negative cooperativity in binding was observed at physiological ionic strength in the absence of transportable anions. When binding was measured by a centrifugation method that did not depend on fluorescence, similar negative cooperativity was observed with stoichiometry of approximately one mole-bound DBDS per mole Band 3. The negative cooperativity and stoichiometry suggest that two DBBS molecules interact in binding to a Band 3 dimer.

Further support for the close proximity of two binding sites comes from the work of Macara and Cantley,[7] who studied the binding of the reversible inhibitor, $H_2(NBD)_2DS$, and its interaction with the irreversible inhibitor, BIDS. Similar to the results for DBDS, $H_2(NBD)_2DS$ exhibited negative cooperativity in binding; binding of $H_2(NBD)_2DS$ was assayed by quenching of intrinsic tryptophan fluorescence. Attempts to measure binding by centrifugation and absorbance of the supernatant were unsuccessful due to the exceptionally high affinity of $H_2(NBD)_2DS$ for Band 3 (dissociation constant 50 nM).

In another set of experiments, Macara and Cantley partially labeled some of the disulfonic stilbene sites with BIDS, then studied the equilibrium affinity of $H_2(NBD)_2DS$ to the remaining sites. When the sites were 80% occupied with BIDS, the affinity of $H_2(NBD)_2DS$ was lowered by an order of magnitude. If the sites were indeed independent, then pretreatment with BIDS would lower the number of sites available to $H_2(NBD)_2DS$, but would not alter the affinity of the free sites. The fact that the affinity decreased by an order of magnitude suggests that the two sites are close enough to interact.

Fluorescence energy transfer from BIDS to $H_2(NBD)_2DS$ was used to estimate the distance between sites. Transfer efficiency was measured by quenching of BIDS fluorescence. In addition to uncertainties in oligomer interactions and in the precise value of κ^2, as described above for the stilbene-maleimide energy transfer experiments, there are complications due to the statistics of labeling. This problem was addressed by measuring the transfer efficiency in the limit of saturating acceptor, so that every donor was adjacent to an acceptor. The range of distances calculated from the measurements was 2.8 to 4.6 nm, which is comparable to the molecular dimensions of BIDS and $H_2(NBD)_2DS$. A similar distance was calculated from energy transfer measurements from BIDS to eosin 5-maleimide; eosin 5-maleimide is thought to bind competitively with disulfonic stilbenes (see below). Since the calculated distance between binding sites is so short, the negative cooperativity observed in DBDS and $H_2(NBD)_2DS$ binding may be due to steric interactions between these rather bulky stilbene inhibitors in binding to monomers of a Band 3 dimer.

Lieberman and Reithmeier[24] studied the equilibrium binding of BADS to red-cell membranes. In contrast to the negative cooperativity observed with DBDS and $H_2(NBD)_2DS$, the binding of BADS was characterized by a single class of sites. Binding was assayed by the fluorescence enhancement method and confirmed by the centrifugation method, using radioactively labeled BADS. A single class of sites for BADS was also found by Tinklepaugh,[25] who used the fluorescence enhancement method. Scatchard plots of binding with radioactively labeled BADS revealed a nonsaturable component which was not observed with DBDS and $H_2(NBD)_2DS$; the reason for this difference is not clear. Since BADS and DBDS differ by just one benzamido group at the end of the molecule, these results suggest that the binding sites on two monomers of a Band 3 dimer are just far apart enough so that two BADS molecules do not interact sterically, but close enough so that two DBDS molecules can interact.

Although BADS and smaller disulfonic stilbenes can probably bind independently to monomers, the dimeric structure seems to be necessary for the integrity of the binding site on Band 3. Band 3 is normally a dimer in solution. Monomeric Band 3 can be prepared by attaching the dimer to a *p*-chloromercuribenzoate Sepharose® 4B affinity column, denaturating the protein to form monomers, then renaturing by removal of the denaturants on the column to form monomeric Band 3 attached to the column. BADS does not bind to this monomeric form of Band 3.[26] BADS binding in the normal 1:1 stoichiometry can be recovered by eluting the monomeric Band 3 from the column with spontaneous dimer reformation. These results may mean that the disulfonic binding site is at or near the interface between two monomers; alternately, the energetic interactions accompanying dimer formation may, through a conformational change in the protein, create the binding site in some other part of the protein.

IV. MECHANISM OF INHIBITOR BINDING

Since several disulfonic stilbenes bind reversibly to Band 3 and undergo fluorescence enhancement, it is possible to measure conveniently the kinetics of binding and, coupled with equilibrium binding studies, to determine a mechanism for inhibitor binding. Once the mechanism is established, then the effects of other compounds on Band 3 can be determined by measuring their effects on the binding mechanism.

The binding mechanism of DBDS to Band 3 has been studied in the most detail.[22,23]

Binding kinetics have been measured by both the temperature-jump method and the stopped-flow method. The buffer chosen for these studies was 28.5 mM sodium citrate, pH 7.4. The ionic strength of this buffer is 160 mM, characteristic of blood serum, and citrate is not transported and interacts only weakly with Band 3.[27] The aim of these experiments was to establish a benchmark against which effects of other compounds (including transportable anions) on Band 3 could be determined. Maintenance of ionic strength is important since disulfonic stilbene binding to Band 3 depends strongly on ionic strength.[7,13,23,24]

Sample temperature-jump and stopped-flow data for DBDS binding to Band 3 are given in Figure 5. The temperature-jump time courses (Figure 5A) follow a single exponential decay of fluorescence, while the stopped-flow time courses (Figure 5B) generally follow a double exponential increase. As in classical chemical kinetics, a minimal reaction mechanism can be determined by measuring the rate of reaction as a function of reactant concentration. Figure 6A shows the inverse exponential time constant from temperature-jump experiments as a function of the sum of unbound DBDS and unbound Band 3; the unbound concentrations were determined by equilibrium-binding experiments described in the previous section. The data exhibit hyperbolic saturation with the sum of free concentrations. The hyperbolic dependence implies that the mechanism consists at least of a rapid bimolecular association of reactants (forward rate constant, k_1, reverse rate constant k_{-1}), followed by a slower unimolecular step (forward and reverse rate constants k_2 and k_{-2}):[28]

$$\text{Band 3} \xrightarrow{k_1} \text{Band 3-DBDS} \underset{k_{-2}}{\overset{k_2}{\rightleftarrows}} \text{Band 3*-DBDS}$$

$$\Big\downarrow k_{-1}$$

$$\text{DBDS} \tag{1}$$

The fast bimolecular-slow unimolecular reaction mechanism gives a hyperbolic dependence of inverse temperature-jump time constant (τ_{TJ}) on sum of free concentrations:

$$1/\tau_{TJ} = k_2([\text{Band 3}] + [\text{DBDS}]/((k_{-1}/k_1) + [\text{Band 3}] + [\text{DBDS}])) + k_{-2} \tag{2}$$

While time courses for perturbation experiments such as temperature-jump will always be one or more exponentials for small enough perturbations, time courses for stopped-flow experiments will not necessarily be exponential since typically, these experiments at zero time are far from equilibrium. A common simplification of the experimental conditions is to use an excess of one reactant so that its concentration becomes time-independent and bimolecular reactions steps become pseudo-first order steps. The original stopped-flow data of Verkman et al.[23] was obtained with a home-built apparatus having a relatively large dead time on the order of 100 msec; consequently, the rapid time course seen in Figure 5A was not seen in these experiments and a single exponential function was fit to the data. For a rapid bimolecular step, followed by a slower unimolecular step, as in Equation 1, the inverse stopped-flow time constant (τ_{SF}) is related to the total DBDS ([DBDS]$_T$) in the limit of excess DBDS, by:

$$1/\tau_{SF} = k_2[\text{DBDS}]_T/((k_{-1}/k_1) + [\text{DBDS}]_T) + k_{-2} \tag{3}$$

If k_{-2} is very small, then a plot of τ_{SF} vs. $1/[\text{DBDS}]_T$ should be linear with slope = $(k_{-1}/k_1)/k_2$ and y-intercept = $1/k_2$. Figure 6B shows that the stopped-flow time constant is indeed linear with inverse [DBDS]$_T$, supporting the minimal binding mechanism given in Equation 1 with k_{-2} small.

In spite of the qualitative agreement between stopped-flow and temperature-jump data,

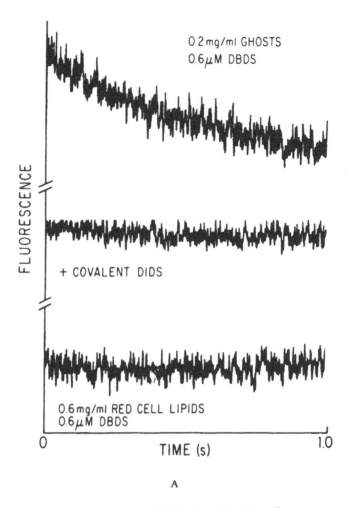

FIGURE 5. Time course of DBDS binding to Band 3. (A) Temperature-jump experiments. In the upper trace, DBDS and isolated red-cell membranes (ghosts) were allowed to reach equilibrium at 23°C, then perturbed by a temperature-jump of 3°C within 6 μsec by discharging a capacitor through the solution. The resulting decrease in fluorescence represents an unbinding of DBDS from Band 3, and is described well by a single exponential function. The middle trace represents the same experiment, except with membranes previously labeled with the irreversible stilbene inhibitor, DIDS. The absence of a time course shows that the temperature-jump fluorescence signal of DBDS arises from unbinding from Band 3. The bottom trace is signal observed from temperature-jump experiments with vesicles made from extracted red cell lipids. (Reproduced from The *Journal of General Physiology*, 1983, 81, 421—449, by copyright permission of the Rockefeller University Press.) (B) Stopped-flow experiments. Red-cell ghost membranes suspended in 28.5 m*M* sodium citrate buffer, pH 7.4, were mixed with 20 μ*M* DBDS in a stopped-flow photometer. The resulting time course is well-described by a double exponential function.[51]

there is a quantitative difference in the value of the equilibrium-binding parameter k_{-1}/k_1 determined from the two sets of data: for stopped-flow, $k_{-1}/k_1 = 3$ μ*M*, while for temperature-jump, k_{-1}/k_1 is about an order of magnitude smaller. In order to explain this discrepancy, Verkman et al.[23] suggested that the bimolecular step was not fast enough to ignore in temperature-jump experiments; in stopped-flow experiments, the bimolecular re-

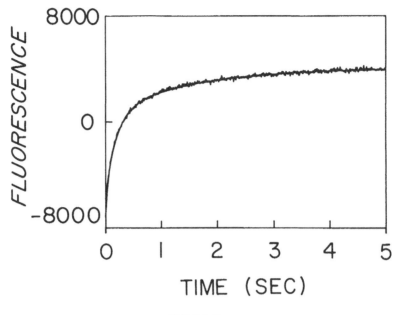

FIGURE 5B.

action was driven to a high rate by the large excess DBDS concentration and was effectively uncoupled with the slower time course. This suggestion was supported by Smith and Dix,[29] who, using a stopped-flow apparatus with a 2 msec dead time, were able to observe a second, fast, exponential time course (Figure 5B). The fast time course corresponded to a bimolecular rate constant of $10^6 M^{-1} sec^{-1}$, of the same order of magnitude as that predicted by Verkman et al. to explain the discrepancy between stopped-flow and temperature-jump data.

The complete mechanism of DBDS binding to Band 3 has to take into account the two classes of binding sites seen in equilibrium binding experiments. Since the stoichiometry of binding is 1:1 and DIDS, which is known to bind predominately to Band 3, completely abolishes DBDS binding as measured by fluorescence enhancement, both classes of sites must be associated with Band 3. Two orthogonal models of binding are an independent site model, in which DBDS can bind to one class of sites independent of binding to the other class, and a negative cooperative model, in which there is a single class of sites, but DBDS binding to one site alters the affinity of an adjacent site. In view of the steric hindrance suggested by the experiments described in the previous section, the negative cooperative model seems more likely. A detailed analysis of the concentration dependence of the temperature-jump time constants supports this hypothesis.[23] Therefore, a minimal model consistent with the equilibrium binding and kinetic data is

$$[Band\ 3] \leftrightarrows [Band\ 3\text{-}DBDS] \longleftrightarrow [Band\ 3^*\text{-}DBDS] \leftrightarrows [Band\ 3^*\text{-}(DBDS)_2]$$

[DBDS]   [DBDS]   (4)

A preliminary kinetic characterization of BADS binding has been done by Tinklepaugh.[25] The results for BADS differ from those for DBDS in that only a single class of sites is observed in equilibrium-binding experiments;[24,25] kinetic experiments reveal that BADS binds a factor of five faster than DBDS. BADS exhibits biphasic kinetics of binding, similar to that seen for DBDS. These results suggest that a minimal mechanism for BADS binding to Band 3 is a bimolecular step followed by a unimolecular step, as given in Equation 1.

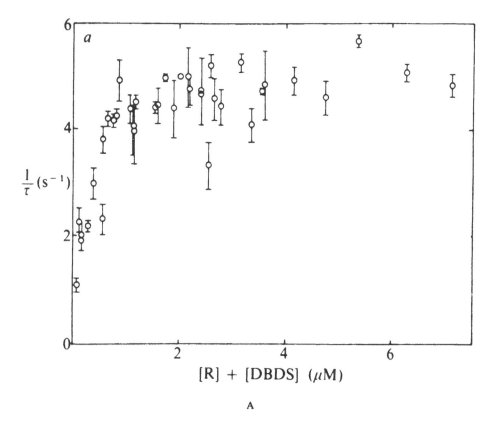

A

FIGURE 6. Concentration dependence of DBDS binding rate. (A) Temperature-jump data. The ordinate gives the reciprocal time constant from a fit of a single exponential function to temperature-jump data as given in Figure 5. The abscissa is the sum of free Band 3 (R) and DBDS concentrations as determined by equilibrium-binding experiments. (Reprinted by permission from *Nature*, 282, 520—522. Copyright (c) 1979 Macmillan Journals Limited.) (B) Stopped-flow data. The ordinate gives the time constant obtained from a fit of a single exponential function to stopped-flow time courses, starting the fit at 120 msec. The abscissa is the inverse of the total DBDS concentration after mixing. The line was fit by the weighted least-squares method.[52]

What is the nature of the first bimolecular step and second unimolecular step in the binding mechanism? A plausible explanation is that the first bimolecular step represents a diffusion- and orientation-controlled association of DBDS with Band 3, and the unimolecular step represents a conformational change in the Band 3 protein induced by DBDS binding. Support for this explanation comes from the temperature dependence of the rate constants for the first two steps.[30] The bimolecular step is characterized by large enthalpy changes and small entropy changes, while the unimolecular step is characterized by both large enthalpy and large entropy changes (Table 2). Similar activation parameters have been reported for enzyme reactions.[31]

That disulfonic stilbenes induce a conformational change in Band 3 is supported by studies using eosin 5-maleimide to probe Band 3. Eosin 5-maleimide has a rather long triplet lifetime of 2 to 3 msec[32] which allows transient absorption dichroism to be observed. Nigg and Cherry[33,34] have used this phenomena to show that the rotational correlation time of Band 3 labeled with eosin 5-maleimide does not change when monomers are cross-linked, sug-gesting that Band 3 exists as a dimer in the native membrane. The binding site for eosin 5-maleimide was originally thought to be in the 42 kdaltons cytoplasmic domain of Band 3, but was subsequently localized on the same 17 kdaltons fragment of the integral membrane

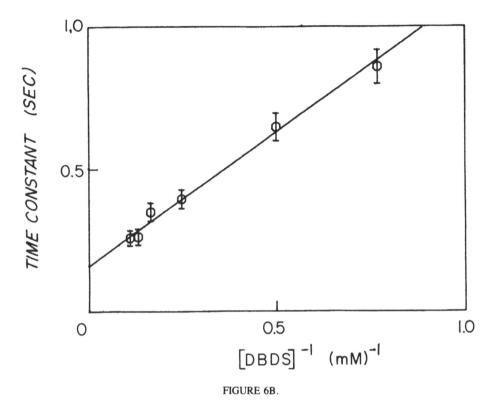

FIGURE 6B.

domain as the H$_2$DIDS binding site.[35] This finding is consistent with the 1:1 relation between the fraction of eosin 5-maleimide bound and fractional inhibition,[34] and the negative linear relation between the amount of BIDS bound and the amount of eosin 5-maleimide bound to Band 3.[7]

Macara et al.[35] have studied the quenching of fluorescence from bound eosin 5-maleimide by CsCl. Although eosin 5-maleimide reacts with Band 3 only when exposed to the extracellular surface of the membrane, extracellular CsCl did not quench the fluorescence of eosin 5-maleimide, whereas CsCl applied to the intracellular surface quenched a maximum of 27% of the fluorescence. The implication is that eosin 5-maleimide may be transported part way through Band 3 to become partially exposed to the cytoplasmic surface. CsCl also quenches the intrinsic tryptophan fluorescence of Band 3; the susceptibility of the tryptophan residues to CsCl quenching was increased when Band 3 was labeled with eosin 5-maleimide or BIDS. These results imply that eosin 5-maleimide and BIDS induce a conformational change that places some Band 3 tryptophan residues in an environment more exposed to CsCl quenching.

V. APPLICATIONS

A. Probe of Band 3 Integrity

The fluorescence assay described above for equilibrium and kinetic binding studies of BADS and DBDS provides a convenient and sensitive assay for the ternary and quaternary structure of Band 3. The assay is convenient because in less than 30 min one can obtain sufficient binding data to determine if Band 3 has been altered with respect to inhibitor binding by some biochemical or biophysical procedure. The assay is sensitive due to the large enhancement in fluorescence upon binding to Band 3.

An example of this use of inhibitor binding is the work of Werner and Reithmeier,[36] who

Table 2
STANDARD THERMODYNAMIC PARAMETERS CHARACTERIZING DBDS BINDING AT 25°C

Parameter	Value	Units	ΔH (kcal/mol)	ΔS (cal/mol · K)	ΔG (kcal/mol)	E_a (kcal/mol)
K_{eq}	0.77	μM	9.1 ± 0.9	3.21 ± 0.3	8.2 ± 0.5	—
k_1	0.67	$sec^{-1} \mu M^{-1}$	11.0 ± 0.9	6.6 ± 0.7	9.0 ± 1.2	11.6 ± 0.9
k_{-1}	2.4	sec^{-1}	17 ± 2	-0.45 ± 0.07	17 ± 2	17 ± 2
k_2	1.31	sec^{-1}	5.7 ± 2.3	-35 ± 18	16 ± 8	6.4 ± 2.3
k_{-2}	0.52	sec^{-1}	10.0 ± 1.9	-26 ± 7	18 ± 4	10.6 ± 1.9

For K_{eq}, ΔH, ΔS, and ΔG represent the equilibrium enthalpy, entropy, and free energy changes, respectively, for the overall DBDS binding mechanism of Equation 1 of the text. For the rate constants, ΔH, ΔS and ΔG represent the activation enthalpy, entropy, and free energy for each individual step in the binding mechanism. E_a represents the activation energy. DBDS binding was measured in 150 mM NaCl, 5 mM HEPES, pH 7.4; under these conditions, a single class of sites is observed for DBDS binding. Standard state is 1 M, activity coefficients equal to 1.

From Posner, R. G. and Dix, J. A., *Biophys. Chem.*, 23, 139, 1985. With permission.

used BADS to probe the structure of Band 3 extracted from membranes with the detergent, octyl glucoside. If Band 3 was extracted with octyl glucoside concentrations below the critical micelle concentration of 40 mM and the detergent concentration subsequently lowered to 2 mM, then the dissociation constant for BADS binding to extracted Band 3 was the same as in the native membrane. However, the BADS dissociation constant increased with increasing detergent concentration present at the time of assay, suggesting that high concentrations of detergent change the tertiary structure of the disulfonic stilbene binding site.

The increase in dissociation constant was reversible as long as the octyl glucoside concentration was below the critical micelle concentration. If, however, Band 3 was extracted with octyl glucoside concentrations above 40 mM, then an irreversible increase of the dissociation constant by an order of magnitude was observed; the dissociation constant of BADS did not return to its native level when the detergent concentration in the assay mixture was lowered. Since extraction and reconstitution studies would typically use a high concentration of detergent, the reconstituted protein may differ significantly from the native protein.

In the above study, the circular dichroism spectrum of Band 3 also showed changes in parallel with changes detected by BADS. However, circular dichroism may not always be sensitive enough to detect structural alterations.[36] For example, if 1% of sodium dodecyl sulfate is added to solubilized Band 3, then no BADS binding can be detected,[24] even though the circular dichroism spectrum reveals relatively little change in the secondary structure.[37]

B. Effect of Transportable Anions
Dix et al.[38] studied the effect of the transportable anion, chloride, on the equilibrium and kinetic properties of DBDS binding to Band 3. In contrast to pure citrate buffer, DBDS binding in 5 mM chloride buffer revealed only a single class of sites. In order to explain this observation, Dix et al. proposed a very high-affinity chloride binding site which, when bound by chloride, changed interacting monomers of a Band 3 dimer into noninteracting monomers.

At constant ionic strength of 160 mM (maintained with sodium citrate), the apparent dissociation constant of DBDS increased with chloride concentration (Figure 7A). Based on effects of disulfonic stilbenes on anion transport,[7,8] the proposed mechanism of inhibition is simple competition, in which an external transport site can be bound by either an anion

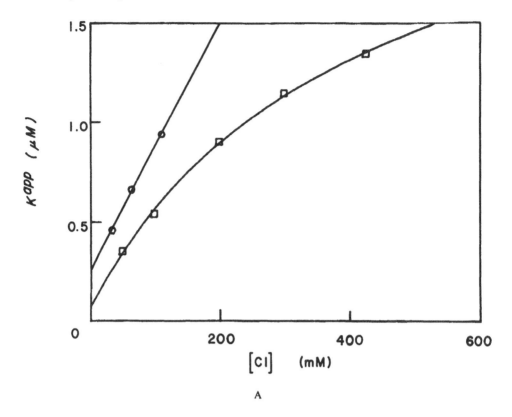

FIGURE 7.   Effect of chloride on DBDS binding. (A) Equilibrium binding. The apparent dissociation constant for DBDS binding to Band 3 is plotted as a function of chloride concentration at constant ionic strength of 160 mM (○) and 600 mM (□). Ionic strength was maintained with sodium citrate. The 160 mM ionic strength data are fit to a linear function; the 600 mM ionic strength data are fit to hyperbola. The −x intercepts of the plots, which give the apparent dissociation constant of chloride from its site, are 40 mM (160 mM ionic strength) and 11 mM (600 mM ionic strength). (From Dix, J. A., Verkman, A. S., and Solomon, A. K., *J. Membr. Biol.*, 89, 211, 1986. With permission.) (B) Binding kinetics. Red-cell ghost membranes were mixed with 35 μM DBDS in a stopped-flow photometer with either 28.5 mM sodium citrate (bottom curve) or 8.5 mM sodium citrate/50 mM sodium chloride (top curve), pH 7.4. A double exponential fit to the data gave time constants of 47 msec and 426 msec for citrate, and 31 msec and 240 msec for 50 mM chloride.[51]

or an inhibitor, but not by both simultaneously (Figure 1). The linear relation between apparent dissociation constant and chloride concentration seen at 160 mM ionic strength supports this view.

However, two observations suggest that the interaction may not be simple competition. If the apparent DBDS dissociation constant is measured as a function of chloride concentration at constant ionic strength of 600 mM, then the relation is no longer linear but saturable (Figure 7A), implying that at infinite chloride concentration, some DBDS could still be bound. The second observation incompatible with simple competition is the kinetic results shown in Figure 7B. If the interaction was competitive, the rate of DBDS binding should decrease with increasing chloride concentration. Contrary to this prediction, the rate of binding increases when chloride is present.

In order to account for these observations, Dix et al.[38] proposed a mechanism in which chloride and DBDS have distinct interacting binding sites (Figure 8) and in which the inhibition is predicted to be a linear mixed-type inhibition. In the absence of chloride, DBDS can bind by the upper pathway of Figure 8. As chloride is added, DBDS can also bind by the lower pathway. In the limit of infinite chloride concentration, DBDS can bind only by

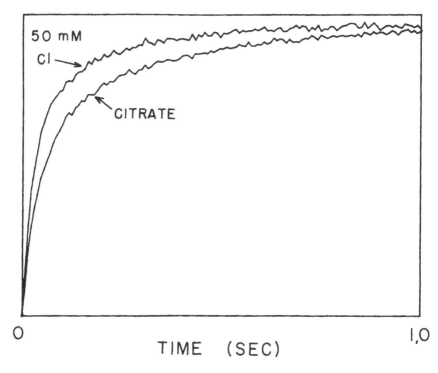

FIGURE 7B.

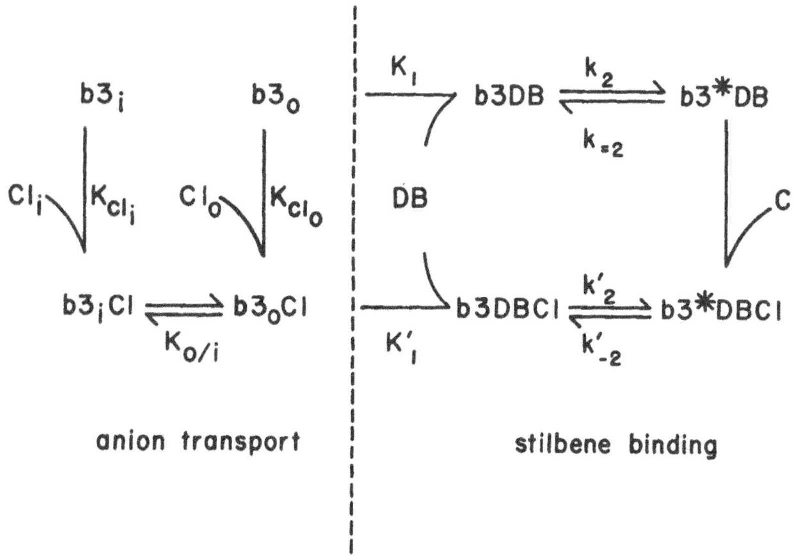

FIGURE 8. Relation between anion transport and DBDS binding. The reaction steps to the left of the vertical dashed line represent the ping-pong mechanism of anion transport as given in Figure 1. The reaction steps to the right of the dashed line represent the DBDS binding steps. If all forms of Band 3 that are bound with DBDS cannot transport anions, as illustrated, then the inhibition would be linear, mixed type, and Dixon plots of inhibition of anion transport would be linear. (From Dix, J. A., Verkman, A. S., and Solomon, A. K., *J. Membr. Biol.*, 89, 211, 1986. With permission.)

the lower pathway, and the apparent DBDS dissociation constant saturates. By making the rate constants for the bottom pathway faster than those of the upper pathway, chloride can increase the rate of DBDS binding; by suitable combination of the ratio of rate constants, chloride can also increase the apparent DBDS dissociation constant.

By extrapolation of the equilibrium binding data of Figure 7B to the x-axis, the apparent dissociation constant for chloride binding to its site can be determined; at 160 mM ionic strength, the dissociation constant is 40 mM, while at 600 mM ionic strength, the dissociation constant drops to 11 mM. These values suggest that chloride binding to Band 3 has an electrostatic contribution; charges may be more shielded at higher ionic strength, allowing chloride to bind more tightly. Precise values for the chloride and DBDS equilibrium constants for the mechanism in Figure 8 are difficult to ascertain because of the uncertainty of the values of the equilibrium constants representing the b3$_o$ to b3$_i$ transition.

Smith and Dix[39] studied the effect of the monovalent anions, fluoride, chloride, bromide, and bicarbonate on DBDS equilibrium binding at 160 mM ionic strength. Attempts to determine the effect of iodide on DBDS binding were hindered by iodide quenching of DBDS fluorescence. They found the affinity to be in the sequence HCO_3 > Br > Cl > F; this selectivity sequence implies that the anion binding site is a weakly cationic site.[40]

C. Effect of Other Compounds

The DBDS-Band 3 interaction has been used to monitor the interaction of the water transport inhibitor, *p*-chloromercuribenzene sulfonate (pCMBS),[41] of the glucose transport inhibitor and noncompetitive anion transport inhibitor, phloretin,[42] and of the anesthetics, halothane and *n*-alkanols,[43] with Band 3. Most of these compounds inhibit DBDS binding in equilibrium experiments (the notable exception is the long-chain alcohols, which do not affect DBDS equilibrium binding); however, in kinetic experiments, some compounds (halothane and *n*-alkanols, as well as transportable anions) accelerate the rate of DBDS binding, while other compounds (phloretin and pCMBS) slow down the rate of binding. These applications are discussed in detail in the next sections.

1. pCMBS

The mechanism and pathways of water transport in red cells are not known with certainty. Some water presumably passes through red cell proteins, since pCMBS, a mercurial reagent known to react with sulfhydryl groups of membrane proteins, inhibits 50 to 60% of the diffusional and 90% of the osmotic water transport in red cells.[44,45] The pathway of the remaining water flux may be through membrane lipids or may be through a protein pathway not inhibitable by pCMBS.[44]

Solomon et al.[46] have suggested that Band 3 is a major transport pathway for water in the red cell. A key piece of evidence for this assertion was the specific interaction found between pCMBS and DBDS.[41] These experiments were done with the cells prelabeled with NEM so that five Band 3 sulfhydryl groups were masked. In spite of NEM prelabeling, the characteristics of water transport inhibition and of the pCMBS-DBDS interaction are not appreciably affected, leading Solomon et al.[46] to define a sixth "cryptic" sulfhydryl group on Band 3, not available for NEM binding.

The effect of pCMBS on DBDS equilibrium binding and stopped-flow kinetics is shown in Figure 9. The apparent DBDS dissociation constant saturates at high pCMBS concentration (Figure 9A), similar to results found for chloride equilibrium binding at 600 mM ionic strength (Figure 7). In contrast to chloride, however, pCMBS decelerates the rate of DBDS binding (Figure 9B). The saturation in the equilibrium-binding constant implies that DBDS and pCMBS have separate but interacting sites. The decrease in DBDS binding rate with increasing pCMBS concentration implies that the activation barriers for DBDS binding are increased.

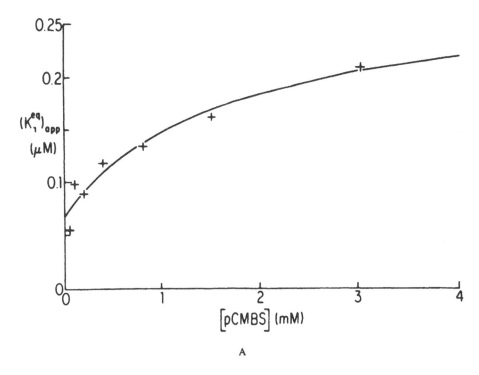

A

FIGURE 9. Effect of pCMBS on DBDS binding. (A) Equilibrium binding. The apparent DBDS dissociation constant determined from fluorescence enhancement is plotted as a function of pCMBS concentration.
For these experiments, membranes had been prelabeled with NEM. The fitted line is a hyperbolic fit with
half saturation at pCMBS concentration of 1.7 mM. (B) Binding kinetics. Red-cell ghost membranes,
prelabeled with NEM, were suspended in 28.5 mM sodium citrate buffer, pH 7.4, with varying amounts
of pCMBS, then mixed with solution containing DBDS and pCMBS. The resulting fluorescence time course
was fit to a single exponential function and the fitted time constant plotted as a function of inverse DBDS
concentration for varying pCMBS concentrations. The lines are the weighted linear least-squares lines.
(From Lukacovic, M. F., Verkman, A. S., Dix, J. A., and Solomon, A. K., *Biochim. Biophys. Acta*, 778,
253, 1984. With permission.)

Studies of water and anion transport inhibition in red cells imply that water and anion
transport are not coupled: DBDS inhibits anion transport but not water transport, and pCMBS
inhibits water transport but not anion transport. However, the data shown in Figure 9 clearly
indicate that the two inhibitors interact. Since DBDS is a probe specific to Band 3, these
data suggest that pCMBS also binds to Band 3. Support for this view comes from radioactive
labeling studies. If cells are treated with NEM, then with [203Hg]pCMBS, most of the
radioactivity is found in Band 3 when the proteins are separated by gel electrophoresis.[46]

More work is needed to establish if Band 3 serves as a protein water transport pathway
in the red cell. If water does indeed move through Band 3, then the observed
pCMBS/DBDS interaction suggests that Band 3 exists as functionally independent domains,
each with its own transport inhibition site. One domain could transport anions while the
other domain could transport water. Inhibition of one domain leads to inhibition of transport
through that domain without inhibition of transport through the other domain; however, the
inhibitor binding to one site modifies the characteristics of inhibitor binding to the other
site.

2. Phloretin

The effects of phloretin on the red cell include inhibition of the transport of glucose,
anions, and small nonelectrolytes[45,47,48] and acceleration of transport through membrane

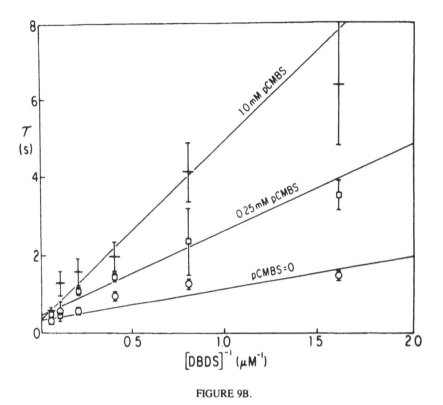

FIGURE 9B.

lipids.[49] The effects of phloretin on the anion transport protein were probed by measuring the effect of phloretin on DBDS equilibrium binding and kinetic parameters.[42] The apparent equilibrium DBDS dissociation constant increased linearly with phloretin concentration, in contrast to that found for chloride and pCMBS; the linear relation suggests phloretin and DBDS compete for a single class of sites on Band 3.

The overall DBDS dissociation constant was resolved into the equilibrium constants for the first and second steps in the DBDS binding of Equation 4; the dependence of the equilibrium constants for these two steps on phloretin concentration is shown in Figure 10A. As would be expected for a competitive interaction, phloretin affects only the bimolecular step and not the unimolecular step in the DBDS binding mechanism.

Phloretin also slows down the kinetics of DBDS binding, as shown in Figure 10B. In this figure, data were obtained at low DBDS concentrations, so that the first term on the right in Equation 3 was not small compared to k_{-2}, and the stopped-flow time constant was not linear with inverse DBDS concentration. Consequently, the full equation was fit to the data and the resulting curves are no longer linear. As expected for a competitive inhibitor, phloretin decreases the binding rate of DBDS. The apparent inhibition constant of phloretin determined from analysis of stopped-flow data was 1.6 μM, in excellent agreement with the value of 1.8 μM determined from the equilibrium-binding experiments. The close agreement of inhibition constants determined from equilibrium and kinetic data strongly supports the idea that phloretin and DBDS interact competitively in binding to Band 3.

Phloretin is a weak acid with pK_a of 7.3. In order to determine whether the charged or uncharged form of phloretin is active in competing with DBDS, the effect of pH on the phloretin-DBDS interaction was studied. Red cell membranes were labeled with DBDS, then titrated with phloretin at a pH below, equal to, and above the pK_a. The resulting decrease in DBDS fluorescence was then plotted as a function of charged phloretin concentration and uncharged phloretin concentration. The data fell along a straight line only

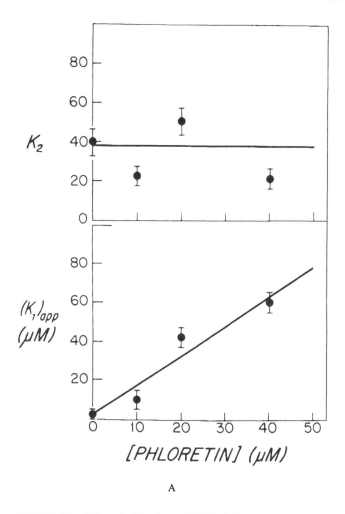

A

FIGURE 10. Effect of phloretin on DBDS binding. (A) Equilibrium binding. The apparent DBDS equilibrium binding constants for DBDS binding in the presence of phoretin were determined, then analyzed in terms of the two-step mechanism of Equation 1 of the text. $(K_1)_{app}$ is the apparent equilibrium dissociation constant of the first bimolecular step in the binding mechanism, and K_2 is the association constant of the second unimolecular step. (B) Binding kinetics. Stopped-flow time constants, determined from kinetic studies, as in Figure 9, are plotted as a function of inverse DBDS concentration for various phloretin concentrations. The lines represent the fit of Equation 2 to the data. (From Forman, S. A., Verkman, A. S., Dix, J. A., and Solomon, A. K., *Biochim. Biophys. Acta*, 689, 531, 1982. With permission.)

when plotted as a function of charged phloretin concentration, indicating that the negatively-charged form of phloretin is active in competition with DBDS. The inhibition constant of phloretin obtained from the pH-dependence data was 1.4 μM, in excellent agreement with that determined by equilibrium and kinetic binding studies. Taken together, these data strongly suggest that the anionic form of phloretin behaves as a DBDS analog in binding to Band 3, as might be expected from the structural similarity of phloretin and DBDS.

3. Anesthetics

The mechanism by which anesthetics exert their effects on biological processes is not

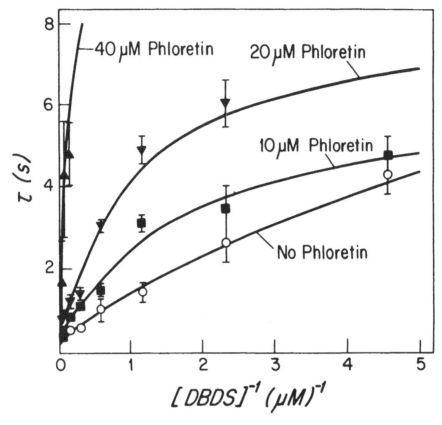

known. Forman et al.[43] used the anion transport system of the red cell as a model to investigate the anesthetic action of halothane and the *n*-alkanols, ethanol, butanol, hexanol, octanol, and decanol. In this study, the effects of the anesthetics on both anion transport and DBDS binding were measured. The anesthetics inhibited anion transport noncompetitively. The inhibition was characterized by a Hill coefficient of 1.3 ± 0.1. Although the IC_{50} values (the concentration of compound at which transport is 50% inhibited) varied widely with total anesthetic concentration, all anesthetics exhibited the same concentration dependence of inhibition when correlated with their membrane lipid concentration, as calculated from partition coefficients. This finding strongly suggests that anesthetics exert their effect on the anion transport protein through the lipid part of the membrane.

In equilibrium-binding experiments, hexanol, octanol, and decanol had no effect on the DBDS dissociation constant, while ethanol and butanol increased the dissociation constant. In kinetic experiments, halothane and the *n*-alkanols increased the DBDS binding rate. When the equlibrium binding and the kinetic data were analyzed by the model of Equation 4, the major effect of the anesthetics was on the forward rate constant for the unimolecular step; the biomolecular step was not affected. This is in contrast with phloretin, a competitive inhibitor of DBDS binding, which affects only the bimolecular step and not the unimolecular step in binding (Figure 10A). As with the transport inhibition data, the effects of anesthetics on DBDS binding correlated with the membrane lipid concentration of the anesthetic, as shown in Figure 11.

The apparent contradiction that, on one hand, anesthetics inhibit anion transport, and, on the other hand, anesthetics accelerate the DBDS binding rate, is explained by postulating that Band 3 can exist in many different conformational states. Some of these states are

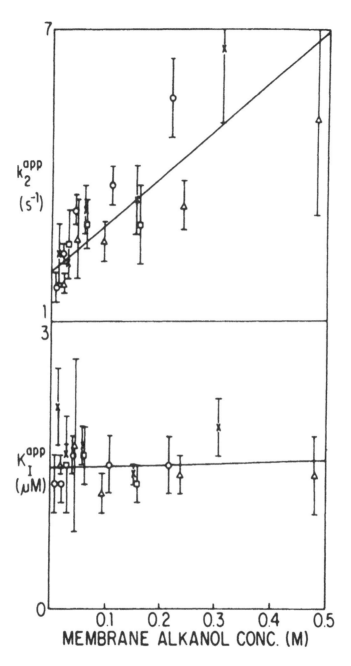

FIGURE 11. Effect of *n*-alkanols on DBDS binding. Stopped-flow kinetic and equilibrium-binding data were obtained for DBDS binding to Band 3 in the presence of various concentrations of *n*-alkanols. The data were analyzed by the model of Equation 1. The lower graph represents the effect of the *n*-alkanols on the apparent equilibrium dissociation constant of the first bimolecular step. The upper graph represents the effect on the forward rate constant for the unimolecular step. The data are plotted as a function of alcohol concentration in the red-cell membrane, as determined from partition coefficients. (Reprinted with permission from Forman, S. A., Verkman, A. S., Dix, J. A., and Solomon, A. K., *Biochemistry*, 24, 4859, 1985. Copyright 1985 American Chemical Society.)

Table 3

EFFECTS OF COMPOUNDS ON DBDS EQUILIBRIUM BINDING AND BINDING KINETICS

Compound	$(K_1^{app})^a$	Binding rate	$(K_1)^b$	$(k_2)^c$	$(k_{-2})^d$	Implication
Phloretin	Linear increase	Decrease	Increase	No effect	No effect	Competitive, mutually exclusive binding, increased activation barriers
pCMBS	Saturable increase	Decrease	Increase	No effect	No effect	Noncompetitive, separate binding sites, increased activation barriers
Chloride	Saturable increase	Increase	Increase	Increase	Increase	Mixed competitive, separate binding sites, lowered activation barriers
Hexanol, octanol, decanol	No effect	Increase	No effect	Increase	Increase	Noncompetitive, separate binding sites, lowered activation barriers
Ethanol, butanol	Increase	Increase	Increase	Increase	—	Noncompetitive or mixed competitive, separate binding sites, lowered activation barriers

[a]   Apparent DBDS equilibrium dissociation constant. For compounds other than chloride, DBDS binds to two classes of sites on the red-cell membrane, according to Equation 4 of the text. The effect given in this column is the effect on the overall equilibrium constant of the first step. In solutions containing chloride, only a single class of sites is observed for DBDS binding; for chloride, then, the effect given in the column is the effect on the single equilibrium constant.

[b]   Equilibrium constant of the first bimolecular step in binding, as given in Equation 4 of the text (or Figure 8 in the case of chloride).

[c]   Forward rate constant of the unimolecular step in DBDS binding.

[d]   Reverse rate constant of the unimolecular step in DBDS binding.

"functional"; i.e., some states can transport anions. Anesthetics, by partitioning into membrane lipids, lower the activation barriers for the protein to adopt nonfunctional states, thus leading to transport inhibition. The long-chain *n*-alkanols do not alter the DBDS binding site topology as probed by the first bimolecular step in binding, but accelerate the unimolecular step, representing a more easily accessible nonfunctional conformation of Band 3 induced by DBDS.

A summary of the effects of chloride, pDMBS, phloretin, and anesthetics on DBDS binding is given in Table 3. The second column gives the effect of the compound on the apparent DBDS equilibrium dissociation constant and the third column gives the effect seen in kinetic experiments. The fourth, fifth, and sixth columns give the effect on the equilibrium constant of the first bimolecular step in binding, the forward rate constant for the unimolecular step, and the reverse rate constant for the unimolecular step, respectively, when the equilibrium binding and kinetic data are analyzed according to the reaction mechanisms given in Equations 1 and 3, and Figure 8.

Several different modes of interaction can be identified in Table 3. For a purely competitive interaction, as is the case for phloretin, K_1 increases while there is no change in k_2 and k_{-2}. Interactions that are not competitive are characterized by a saturable increase in the apparent equilibrium dissociation constant (for chloride and pCMBS) and/or an increase in the binding rate (for chloride and anesthetics). Interactions that are not competitive imply a binding site distinct from the disulfonic stilbene binding site. The binding site may also be on Band 3, as is presumably the case for chloride, since it is transported by Band 3, and possibly for pCMBS as well, since it reacts with membrane proteins. Alternatively, the binding site may be membrane lipids, as is the case for anesthetics, since their effects correlate with their lipid membrane concentration.

VI. CONCLUSIONS

The anion transport system of the human red blood cell has been studied for over 100 years. The results of this study have been a good understanding of the physiological importance of anion transport, the identification of the protein responsible for anion transport, the overall mechanism of transport, and substrate and inhibitor specificity. What is still lacking, however, is a detailed molecular description of the transport process.

The studies described in this chapter represent the use of fluorescent inhibitors of anion transport as a step in obtaining such a molecular description. Fluorescence energy transfer measurements show that the binding site for disulfonic stilbene inhibitors on Band 3 is probably located some distance in from the extracellular surface. The fluorescence properties of bound inhibitor indicate that the binding site is hydrophobic and holds the inhibitor rigidly. Several studies indicate that there is an interaction between monomers of a Band 3 multimer: fluorescence self-energy transfer between FMA, negative cooperativity in $H_2(NBD)_2DS$ and DBDS binding, energy transfer between BIDS and $H_2(NBD)_2DS$, and the requirement of dimeric Band 3 for binding of BADS.

Bound disulfonic stilbene can be used to monitor the integrity and the conformational state of Band 3 subsequent to biochemical manipulation, such as extraction and reconstitution. Binding studies using DBDS to monitor the effect of transportable anions on Band 3 indicate that the anion binding is weakly cationic, and suggest that anion transport occurs through a conformational change in Band 3 that can be blocked by inhibitor. Anesthetics exert their effects on Band 3 through red cell lipids, perhaps by lowering the activation barriers leading to nonfunctional states of Band 3. The anion transport inhibitor, phloretin, appears to bind to the DBDS binding site, while the water transport inhibitor, pCMBS, appears to bind to another site on or near Band 3. The picture that emerges from these studies is that Band 3 is a dynamic protein that can adopt many different conformations; some of these conformations are functional and can transport anions, but access to these states is restricted when inhibitors are bound to Band 3.

The methodologies described in this chapter should prove generally useful in studying ligand binding to macromolecules in other systems. For instance, anion transport plays an important role in the function of mammalian kidney, and specific inhibition of anion transport by disulfonic stilbenes and their analogs has been demonstrated in the proximal straight tubule of rabbit kidney.[50] It should be possible to obtain mechanistic information about inhibitor binding and anion transport in kidney systems using methods described above for red cells.

ACKNOWLEDGMENTS

The work reported from the author's laboratory was supported by NIH grant HL29488. I would like to thank Mr. Kevin Smith, Ms. Sarah Finnegan and Mr. Mark Tinklepaugh for help in the understanding of DBDS and BADS interactions with Band 3. I would also like to acknowledge the assistance of Ms. Jeanne LaBonte and Ms. Maria Perotto in the preparation of this manuscript.

REFERENCES

1. **Knauf, P. A.,** Erythrocyte anion exchange and the band 3 protein: transport kinetics and molecular structure, *Curr. Top. Membr. Transp.,* 12, 249, 1979.
2. **Jennings, M. L.,** Kinetics and mechanism of anion transport in red blood cells, *Annu. Rev. Physiol.,* 47, 519, 1985.

3. **Segel, I.**, *Enzyme Kinetics*, John Wiley & Sons, New York, 1975.
4. **Grinstein, S., McCulloch, L., and Rothstein, A.**, Transmembrane effects of inhibitors of anion transport in red blood cells. Evidence for mobile transport sites, *J. Gen. Physiol.*, 73, 493, 1979.
5. **Furuya, W., Tarshis, T., Law, F.-Y., and Knauf, P. A.**, Transmembrane effects of intracellular chloride on the inhibitory potency of extracellular H₂DIDS. Evidence for two conformations of the transport site of the human erythrocyte anion exchange protein, *J. Gen. Physiol.*, 83, 657, 1984.
6. **Jennings, M. L.**, Oligomeric structure and the anion transport function of human erythrocyte band 3 protein, *J. Membr. Biol.*, 80, 105, 1984.
7. **Macara, I. G. and Cantley, L. C.**, Interactions between transport inhibitors at the anion binding sites of the band 3 dimer, *Biochemistry*, 20, 5096, 1981.
8. **Ship, S., Shami, T., Breuer, W., and Rothstein, A.**, Synthesis of tritiated 4,4'-diisothiocyano-2,2'-disulfonic stilbene disulfonic acid and its covalent reaction with sites related to anion transport in human red blood cells, *J. Membr. Biol.*, 33, 311, 1977.
9. **Kotaki, A., Naoi, M., and Yagi, K.**, A diaminostilbene dye as a hydrophobic probe for proteins, *Biochim. Biophys. Acta*, 229, 547, 1971.
10. **Cabantchik, Z. I. and Rothstein, A.**, The nature of the membrane sites controlling anion permeability of human red blood cells as determined by studies with disulfonic stilbene derivatives, *J. Membr. Biol.*, 10, 311, 1972.
11. **Barzilay, M., Ship, S., and Cabantchik, Z. I.**, Anion transport in red blood cells. I. Chemical properties of anion recognition sites as revealed by structure-activity relationships of aromatic sulfonic acids, *Membr. Biochem.*, 2, 227, 1979.
12. **Hansch, C.**, Quantitative approaches to biochemical structure activity relationships, in *International Encyclopedia of Pharmacology and Therapeutics*, Sect. S, Vol. 1, Pergamon Press, Elmsford, New York, 1973, 75.
13. **Rao, A., Martin, P., Reithmeier, R. A. F., and Cantley, L. C.**, Location of the stilbenedisulfonate binding site of the human erythrocyte anion-exchange system by resonance energy transfer, *Biochemistry*, 18, 4505, 1979.
14. **Lakowicz, J. R.**, *Principles of Fluorescence Spectroscopy*, Plenum Press, New York, 1983, chaps. 7 and 8.
15. **Steck, T. L., Koziarz, J. J., Singh, M. K., Reddy, G., and Kohler, H.**, Preparation and analysis of seven major, topographically defined fragments of band 3, the predominant transmembrane polypeptide of human erythrocyte membranes, *Biochemistry*, 17, 1216, 1978.
16. **Fukuda, M., Eshdat, Y., Tarone, G., and Marchesi, V. T.**, Isolation and characterization of peptides derived from the cytoplasmic segment of band 3, the predominant intrinsic membrane protein of the human erythrocyte, *J. Biol. Chem.*, 253, 2419, 1978.
17. **Ramjeesingh, M., Gaarn, A., and Rothstein, A.**, The location of a disulfonic stilbene binding site on band 3, the anion transport protein of the red blood cell membrane, *Biochim. Biophys. Acta*, 599, 127, 1980.
18. **Rao, A. and Reithmeier, R. A. F.**, Reactive sulfhydryl groups of the band 3 polypeptide from human erythrocyte membranes. Location in the primary structure, *J. Biol. Chem.*, 254, 6144, 1979.
19. **Rao, A.**, Disposition of the band 3 polypeptide in the human erythrocyte membrane. The reactive sulfhydryl groups, *J. Biol. Chem.*, 254, 3503, 1979.
20. **Dale, R. E., Eisinger, J., and Blumberg, W. E.**, The orientational freedom of molecular probes. The orientation factor in intramolecular energy transfer, *Biophys. J.*, 26, 161, 1979.
21. **Dissing, S., Jesaitis, A. J., and Fortes, P. A. G.**, Fluorescence labeling of the human erythrocyte anion transport system. Subunit structure studied with energy transfer, *Biochim. Biophys. Acta*, 553, 66, 1979.
22. **Dix, J. A., Verkman, A. S., Solomon, A. K., and Cantley, L. C.**, Human erythrocyte anion exchange site characterized using a fluorescent probe, *Nature (London)*, 282, 520, 1979.
23. **Verkman, A. S., Dix, J. A., and Solomon, A. K.**, Anion transport inhibitor binding to band 3 in red blood cell membranes, *J. Gen. Physiol.*, 81, 421, 1983.
24. **Lieberman, D. M. and Reithmeier, R. A. F.**, Characterization of the stilbenedisulfonate binding site of the band 3 polypeptide of human erythrocyte membranes, *Biochemistry*, 22, 4028, 1983.
25. **Tinklepaugh, M. S.**, Interaction of Water and Anion Transport Inhibitors with the Red Cell Membrane, M.A. thesis, State Univesity of New York, Binghamton, New York, 1983.
26. **Boodhoo, A. and Reithmeier, R. A. F.**, Characterization of matrix-bound band 3, the anion transport protein from human erythrocyte membranes, *J. Biol. Chem.*, 259, 785, 1984.
27. **Schnell, K. F., Besl, E., and Manz, A.**, Asymmetry of the chloride transport system in human erythrocyte ghosts, *Pfluegers Arch.*, 375, 87, 1978.
28. **Czerlinski, G. H.**, *Chemical Relaxation*, Marcel Dekker, New York, 1966.
29. **Smith, K. R. and Dix, J. A.**, Biexponential kinetics of disulfonic stilbene binding to human red cell ghost membranes, *Biophys. J.*, 45, 202a, 1984.

30. **Posner, R. G. and Dix, J. A.**, Temperature dependence of anion transport inhibitor binding to human red cell membranes, *Biophys. Chem.*, 23, 139, 1985.
31. **Laidler, K. J. and Peterman, B. F.**, Temperature effects in enzyme kinetics, *Methods Enzymol.*, 63, 234, 1979.
32. **Cherry, R. J.**, Measurement of protein rotational diffusion in membranes by flash photolysis, *Methods Enzymol.*, 54, 47, 1978.
33. **Nigg, E. and Cherry, R. J.**, Dimeric association of band 3 in the erythrocyte membrane demonstrated by protein diffusion measurements, *Nature (London)*, 277, 493, 1979.
34. **Nigg, E. and Cherry, R. J.**, Influence of temperature and cholesterol on the rotational diffusion of band 3 in the human erythrocyte membrane, *Biochemistry*, 18, 3457, 1979.
35. **Macara, I. G., Kuo, S., and Cantley, L. C.**, Evidence that inhibitors of anion exchange induce a transmembrane conformational change in band 3, *J. Biol. Chem.*, 258, 1785, 1983.
36. **Werner, P. K. and Reithmeier, R. A. F.**, Molecular Characterization of the human erythrocyte anion transport protein in octyl glucoside, *Biochemistry*, 24, 6375, 1985.
37. **Oikawa, K., Lieberman, D. M., and Reithmeier, R. A. F.**, Conformation and stability of the anion transport protein of human erythrocyte membranes, *Biochemistry*, 24, 2843, 1985.
38. **Dix, J. A., Verkman, A. S., and Solomon, A. K.**, Binding of chloride and a disulfonic stilbene transport inhibitor to red cell band 3, *J. Membr. Biol.*, 89, 211, 1986.
39. **Smith, K. R. and Dix, J. A.**, Interaction between monovalent anions and a disulfonic stilbene in binding to the red cell membrane, *Biophys. J.*, 41, 200a, 1983.
40. **Diamond, J. M. and Wright, E. M.**, Biological membranes: the physical basis of ion and nonelectrolyte selectivity, *Annu. Rev. Physiol.*, 31, 581, 1969.
41. **Lukacovic, M. F., Verkman, A. S., Dix, J. A., and Solomon, A. K.**, Specific interaction of the water transport inhibitor, pCMBS, with band 3 in red blood cell membranes, *Biochim. Biophys. Acta*, 778, 253, 1984.
42. **Forman, S. A., Verkman, A. S., Dix, J. A., and Solomon, A. K.**, Interaction of phloretin with the anion transport protein of the red blood cell membrane, *Biochim. Biophys. Acta*, 689, 531, 1982.
43. **Forman, S. A., Verkman, A. S., Dix, J. A., and Solomon, A. K.**, n-Alkanols and halothane inhibit red cell anion transport and increase band 3 conformational change rate, *Biochemistry*, 24, 4859, 1985.
44. **Dix, J. A. and Solomon, A. K.**, Role of membrane proteins and lipids in water diffusion across red cell membranes, *Biochim. Biophys. Acta*, 773, 219, 1984.
45. **Macey, R. I. and Farmer, R. E. L.**, Inhibition of water and solute permeability in human red cells, *Biochim. Biophys. Acta*, 211, 104, 1970.
46. **Solomon, A. K., Chasan, B., Dix, J. A., Lukacovic, M. F., Toon, M. R., and Verkman, A. S.**, The aqueous pore in the red cell membrane: band 3 as a channel for anions, cations, nonelectrolytes, and water, *Ann. N.Y. Acad. Sci.*, 414, 97, 1983.
47. **LeFevre, P. G. and Mashall, J. K.**, The attachment of phloretin and analogues to human erythrocytes in connection with inhibition of sugar transport, *J. Biol. Chem.*, 234, 3022, 1959.
48. **Krupta, R. M.**, Evidence for a carrier conformational change associated with sugar transport in erythrocytes, *Biochemistry*, 10, 1143, 1971.
49. **Owen, J. D. and Solomon, A. K.**, Control of nonelectrolyte permeability in red cells, *Biochim. Biophys. Acta*, 290, 414, 1972.
50. **Baigi, B. A.**, Effects of the anion transport inhibitor, SITS, on the proximal straight tubule of rabbit perfused in vitro, *J. Membr. Biol.*, 88, 25, 1985.
51. **Smith, K. R. and Dix, J. A.**, unpublished data.
52. **Finnegan, S. and Dix, J. A.**, unpublished data.

Chapter 14

HOW TO CHOOSE A POTENTIOMETRIC MEMBRANE PROBE

Leslie M. Loew

TABLE OF CONTENTS

I. INTRODUCTION

A recent computer search of the Biological Abstracts database discovered over 1000 papers published in the past 7 years which included descriptors linking membrane potential with fluorescence, or dye, or probe. Because of the limited nature of the database and the possibility that these descriptors weren't included in many relevant abstracts, this is easily a 50% underestimate of the actual number of papers published during this period which employed potentiometric spectroscopic probes.

L. B. Cohen and his former co-workers, A. Waggoner, B. M. Salzberg, and A. Grinvald have arguably been responsible for originating and promoting the growth of the probe methodology for following changes in membrane potential. Their approach to probe development has been to screen all available dyes (over 1000 to date) for potential dependent responses to fast-voltage clamp pulses on the squid axon and attempt to obtain improvements by selective synthetic modification of those dyes that appear promising.[1,2,3] In addition, Waggoner demonstrated the extraordinarily high sensitivities of cyanine dyes for slow potential changes in cells in suspension.[4,5]

The approach in this lab has been to synthesize probe structures after a theoretical analysis of possible mechanisms. Electrochromism was the original target because it lends itself to theory[6,7] and we have now synthesized[8] and tested[9,10] about 30 of the resulting "charge-shift probes". More recently, we have tested the idea of measuring membrane potential with cationic permeant dyes which will partition into cells according to the Nernst equation.[11,12]

All this activity has created a vast library of probes available to the physiologist. The rest of this book contains chapters detailing or reviewing the wide range of applications for this technology. The purpose of this chapter is to provide guidelines to aid the uninitiated in selecting the best probe for a particular application. To achieve this, the chemical and spectroscopic variables which influence the usefulness of a probe will be defined, along with those involved in the potential dependent spectroscopic response. It is also hoped that the discussion which follows will help to sharpen the focus on the need for new potentiometric probe development. A literature review of the field is not the intention of this chapter and many important citations have been omitted for the sake of brevity and focus. Most of these are included, however, in the subsequent chapters.

II. SURVEY OF ESTABLISHED MOLECULAR MECHANISMS

A description of the molecular mechanisms by which probes respond to a change in membrane potential will provide a convenient framework for the discussion of response characteristics such as sensitivity, calibratibility, and speed. The mechanisms for many popular dyes are still unknown, but for the cases that have been studied (see Reference 13 for an early but thorough review), only three major types suffice to account for the data. Within each there may be several distinct variations. Also, a given dye may respond via different mechanisms in different membrane, cell, or tissue preparations and may actually adopt several mechanisms concurrently.

A. Redistribution

A hydrophobic ionic dye will have some equilibrium distribution between the external aqueous medium, the cell plasma membrane, the cytosol, and the membranes and aqueous compartments of intracellular organelles. The membrane potential plays a direct role in governing the distribution across and within the plasma membrane — the more negative the potential the greater the accumulation of positively charged dye. All these are coupled equilibria, so that the amount of dye which can associate with the organelles is also indirectly controlled. Thus, the membrane potential controls the averaged environment of the dye;

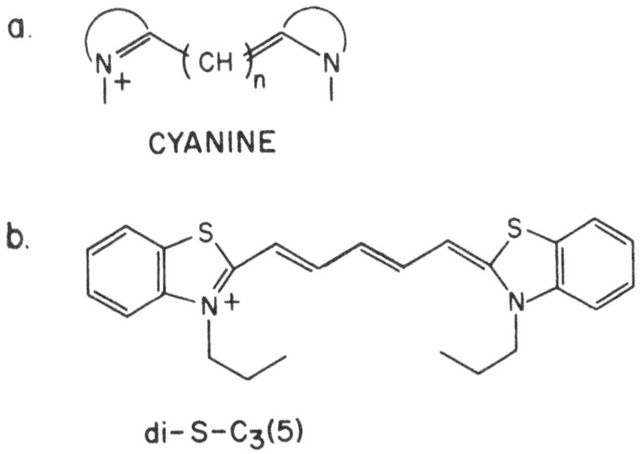

a.

CYANINE

b.

di-S-C₃(5)

FIGURE 1. Cyanine dyes. Only one of two equivalent resonance structures is shown. (a) Generalized structure. (b) di-S-C$_3$(5).

since the fluorescence of a dye is typically very sensitive to its molecular environment, a change in membrane potential can be reflected in a change in fluorescence. The size of this change will depend on the sensitivity of the dye fluorescence to its environment and on the sensitivity of the dye distribution to membrane potential.

Very often dye binding to membranes causes an increase in fluorescence quantum yield and a shift in the wavelength of maximum excitation and emission. If the equilibrium constant for dye binding is such that approximately half is in the membrane and half is in the aqueous medium, a small change in membrane potential will produce the largest possible shift in concentrations of bound vs. free dye; this situation is also the one to give the most linear fluorescence response to the potential variation. This condition, and/or optimal sensitivity, can be achieved by manipulating the concentrations of dye and membrane and by choosing the combination of excitation and emission wavelengths which give the largest relative change in fluorescence. There are several positive cyanine[4] and negative oxonol[14] probes (Figures 1 and 2, respectively) which appear to have adopted this variant of the redistribution mechanism.

Even more dramatic changes in fluorescence are obtained when the dye has a tendency to form nonfluorescent aggregates at high concentration. This phenomenon is commonly observed with positively charged cyanine dyes which accumulate under the influence of negative membrane potentials.[4] The effect is especially pronounced with the more highly conjugated cyanines at higher concentrations. Aggregates can form in either the plasma membrane or cytosol, and, by the law of mass action, their formation drives even more monomer into association with the cell. In a lipid vesicle system, a -170 mV potential caused a 1 million-fold concentration of a cyanine dye with a resultant 98% quench of the fluorescence from the total suspension.[15]

If the ion is not too hydrophobic and if there are no specific binding interactions with protein or nucleic acid molecules in the cell, the distribution will be primarily governed by the Nernst equation:

$$C_{in}/C_{out} = e^{-(V_m ZF/RT)} \tag{1}$$

where V_m is the membrane potential, Z is the charge on the ion, F is Faraday's constant, R is the ideal gas law constant, T is the absolute temperature, and C_{in} and C_{out} are the concentrations inside and outside the cell, respectively, of the permeant ion. This idea is the basis of the radio-labeled tetraphenyl phosphonium (TPP) assay developed by Kaback and his co-workers for measuring membrane potential.[16] Recently, this laboratory has de-

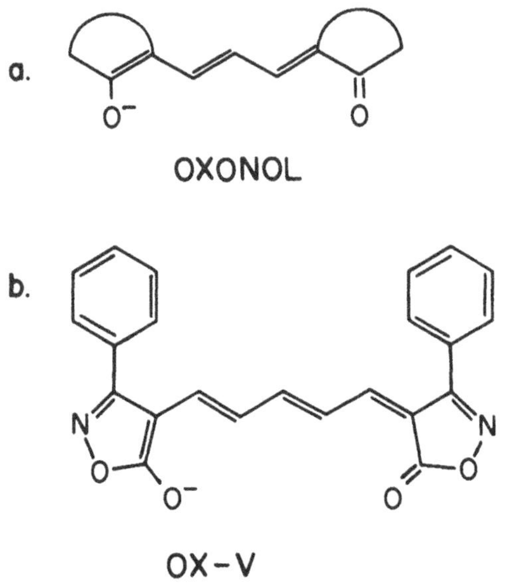

FIGURE 2. Oxonol dyes. Only one of two equivalent resonance structures is shown. (a) Generalized structure. (b) OX-V.

veloped an approach using both existing and custom-synthesized fluorescent cationic dyes which are designed to distribute according to the Nernst equation.[11,12] The compact structures of these dyes, mainly rhodamine analogs (Figure 3), are similar to that of the TPP cation in that there are no long hydrophobic groups to intercallate into the lipid bilayer and there is very little tendency for them to aggregate. Thus, these probes can be used to monitor the potential on individual cells with a microfluorometer; as opposed to the other redistribution-based mechanisms, the total fluorescence from a suspension of cells containing one of these "Nernstian dyes" is not sensitive to the potential. In fact, after correction for some unavoidable background staining, the ratio of fluorescence determined from the inside and outside of a single cell is proportional to the ratio of concentrations in Equation 1 and can therefore be used to determine the absolute value of V_m. The assay also contrasts with the TPP method in that an individual cell may be monitored.

B. Reorientation

A dipolar probe structure may adopt a variety of orientations with respect to the two-dimensional surface of the membrane to which it is bound. The electrostatic interaction of the dipole with a transmembrane electric field can control the relative population of these orientations.

Just as in the case of the redistribution mechanism, there are several ways in which the reorientation of a dye can result in a fluorescence change. The simplest is again based on the sensitivity of fluorescence to the molecular environment of the chromophore; a dye with one end of its long axis buried in the hydrophobic region of the membrane and the other exposed to the aqueous interface is likely to have a different absorption or fluorescence spectrum than the same dye oriented parallel to the membrane surface with both ends experiencing the same average environment. Also, a reorientation-dependent dimerization mechanism has been postulated[2] to account for the action spectra of a merocyanine dye (Figure 4) on the voltage-clamped squid axon. This dye, M-540, was one of the first to display really large voltage-dependent signals and has therefore been the subject of some very thorough mechanistic investigation.[17,18] The importance of a reorientation process for

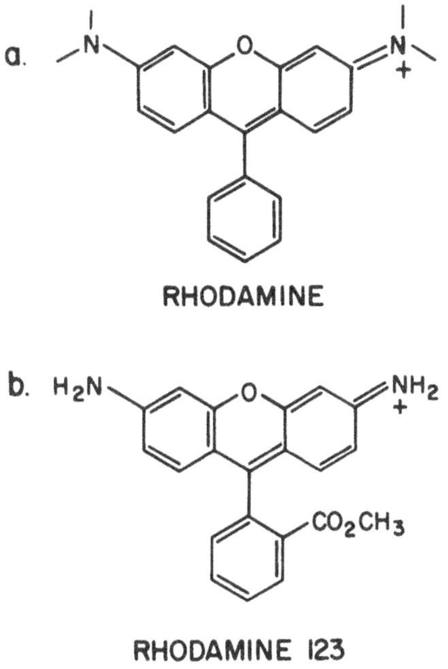

RHODAMINE

RHODAMINE 123

FIGURE 3. Rhodamine dyes. Only one of two equivalent resonance structures is shown. (a) Generalized structure. (b) Rhodamine 123.

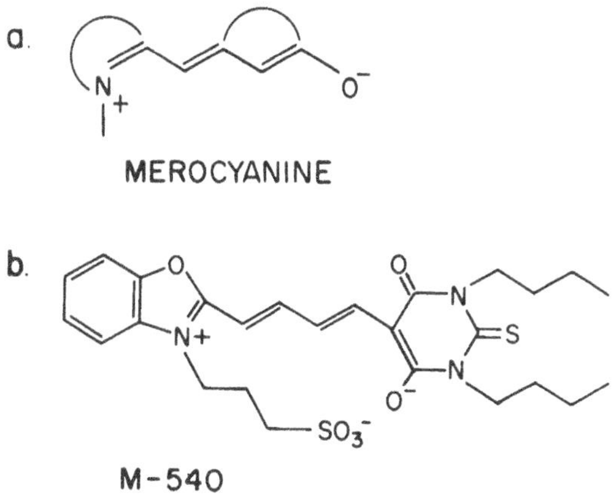

MEROCYANINE

M-540

FIGURE 4. Merocyanine dyes. The dipolar resonance structure is shown; a neutral structure is also a strong contributor to the ground state of these chromophores. (a) Generalized structure. (b) M-540.

M-540 is unquestioned, but there is some controversy with regard to the dimerization component of the mechanism (see the chapter by Salama).

In addition to the above variants of the reorientation mechanism, which can produce voltage-dependent responses for both cells in suspension and individual cells, Dragsten and Webb[17] have detailed the way in which the anisotropy of the fluorescence collected from a membrane patch on an individual cell can lead to a potentiometric signal. The size and polarity of this signal is highly dependent on the geometry of the excitation and emission optics with respect to the membrane patch and can be maximized by employing polarizing filters.

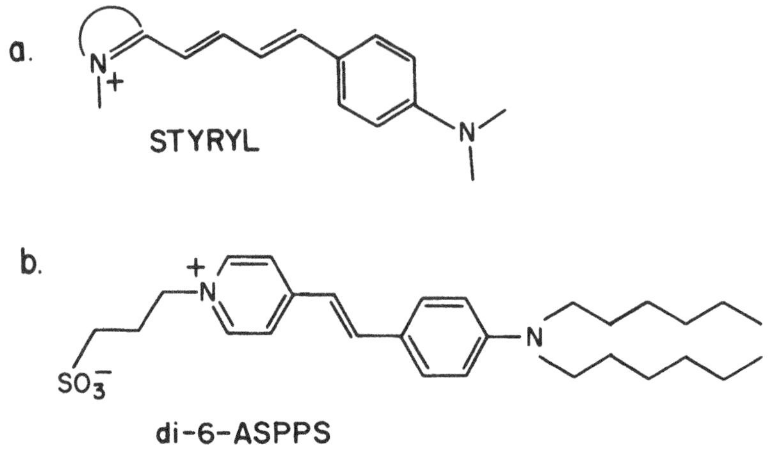

FIGURE 5. Styryl dyes. The structure shown is the primary contributor to the ground state, but the positive charge is partially delocalized. (a) Generalized structure. (b) di-6-ASPPS.

C. Electrochromism

Electrochromism is the direct coupling of the membrane electric field with the electronic redistribution in a chromophore that accompanies excitation or emission.[19] The theory is quite straightforward and can be summarized in a simple equation:

$$h\Delta\nu = -q\mathbf{r} \cdot \mathbf{E} - \Delta\mathbf{\mu} \cdot \mathbf{E} - \Delta\alpha|\mathbf{E}|^2$$

The change in energy for the electronic transition (Planck's constant, h, times the change in light frequency, $\Delta\nu$) upon application of an electric field, $\mathbf{E}$,* depends on the displacement, $\mathbf{r}$, of charge, q, or the difference in dipole moment, $\Delta\mathbf{u}$, or the difference in the polarizability, $\Delta\alpha$, in the two electronic states involved in the transition. The last term, which is manifested as a quadratic dependence on the electric field, is insignificant for fields in the range of interest for physiology. The first two terms can be large enough, however, to allow the design of potentiometric probes.[7] A number of *p*-aminostyryl dyes (Figure 5) display potentiometric responses which are most consistent with an electrochromic mechanism.[9,10]

III. ATTRIBUTES TO BE CONSIDERED IN CHOOSING A PROBE

A. Response Characteristics

The potentiometric response of a dye has attributes which are largely governed by the particular mechanism which couples the membrane potential to the dye spectrum. The size of the spectral change (i.e., the sensitivity of the dye) is largest for the redistribution/aggregation mechanism where 90% changes in fluorescence intensity per 100 mV can be achieved. Sensitivity is often the most important attribute to consider in choosing a dye, but in many applications it must be sacrificed in a trade-off for response speed; the spectral changes of redistribution dyes lag behind voltage steps by times as long as minutes, depending on the particular biological preparation and dye. Clearly, the redistribution dyes are not capable of measuring fast signals such as action potentials.

The fastest response can be expected from the electrochromic mechanism which has the speed of an electronic transition. The size of an electrochromic spectral change has never exceeded about 10%/100 mV.[10] While this is adequate for many applications, fast dyes that

---

* Vectors are indicated by boldface type.

have over twice this sensitivity in particular applications have been described.[20] It has been our experience, however, that these large nonelectrochromic fast signals are very sensitive to the particular biological preparation and to the optical characteristics of a particular experimental apparatus. Often, experimental designs are possible with fast signals which allow for repetitive triggering so that signal averaging can enhance the signal to noise ratio.

The spectral change may be monitored in either fluorescence or absorption modes. There are dyes that respond best in only one of these configurations, however. If the mechanism involves dye aggregation, for example, the spectra of the monomer and dimer are likely to be quite distinct in both absorption and emission; wavelength shifts in the spectra are generally insignificant compared to the change in fluorescence quantum yield for a voltage-dependent aggregation, however. Another factor that favors fluorescence is that the dye concentration can generally be lower than in an absorption experiment. This is because the entire light signal originates from the dye emission, whereas, in the absorbance mode, the light not absorbed by the dye is what is actually being detected. Furthermore, since fluorescence from membrane-bound dye is frequently easily differentiated from that of aqueous dye by the appropriate choice of wavelengths, one can readily exclude background signal levels in the fluorescence mode. On the other hand, Cohen and his co-workers have pointed out that when the fluorescence intensity from the active membrane is low (as in measurements on single axons), shot noise in the photon flux can be the limiting factor in recording a response; under these circumstances, the absorbance mode can give better signal to noise even if the fractional change of the light signal is several orders of magnitude lower than the fractional change in fluorescence for a given voltage change.[2,3] Also, Chance and his co-workers have employed a dual wavelength spectrophotometer to enhance the sensitivity of absorbance measurements for cells in suspension.[14] Obviously, dyes that do not have a significant fluorescence efficiency can only be used in the absorbance mode (note the speculation, in Chapter 22 by Ehrenberg, on the possibility that the Raman spectra of nonfluorescent dyes may be sensitive to potential).

B. Molecular Properties of Dyes

Large extinction coefficients (10^5 OD $\cdot$ cm^{-1} M^{-1}) and fluorescence quantum yields (0.5 to 1) are advantageous for any dye because potential sensitivity can be maintained at low nontoxic dye levels. The toxicity can of course vary from dye to dye and, with a given dye, from cell type to cell type. Somewhat related to toxicity is the ability of the dye to photosensitize the production of singlet oxygen, which can harm cells ("photodynamic damage"). This can be minimized by reducing the incident light intensity, excluding oxygen, or, again, by employing low dye concentrations.

While toxicity and photodynamic effects are somewhat difficult to predict *a priori* for a given molecular structure, other dye attributes which can affect practical applications are predictable, or at least easily determined. The ionic charge of the probe can perturb the surface potential of the target membrane. The emission and absorption spectra can be easily obtained in the presence of lipid viscles to establish if there will be any interference from chromophores intrinsic to the cell or tissue system to be investigated. The fluorescence Stokes shift (i.e., the separation between the excitation and emission spectra) is ideally large for a good probe so as to minimize contributions from scattered light to the fluorescence signal; this is especially critical for turbid or opaque preparations where front-surface illumination and detection may be required. The dark- and photo-stability of a dye can be determined by monitoring its absorption spectrum as a function of time in the presence of an appropriate concentration of vesicles, with or without constant illumination.

For membrane-staining dyes, an important consideration is the strength and speed of binding. Obviously, the more hydrophobic a chromophore or its sidechains, the greater its binding constant to membranes. On the other hand, very hydrophobic dyes will form stable aggregates in water or will not be sufficiently soluble to permit acceptable staining rates. In general, the best compromise appears to be a pair of 3-6 carbon sidechains on a moderately polar chromophore. Also, Cohen and his co-workers[1] have employed a polyol surfactant, which appears to break up dye aggregates without disrupting membranes,[21] to promote staining with insoluble potentiometric probes.

C. Calibration

Unfortunately, no existing potentiometric indicator is universally calibratible; that is, a calibration of spectral change vs. membrane potential which is obtained for a dye in one preparation is not applicable to the same dye in a different preparation. Thus, calibration procedures must be developed for each application of a dye in which a change in fluorescence or absorbance is to be translated into a voltage change; furthermore, it is often impossible to devise a means for determining the resting potential (as opposed to a potential change), especially on optically heterogeneous single cells or tissues. There are, however, combinations of dye mechanism and cell preparation which are amenable to calibration. These will be briefly summarized here.

For cells in suspension measured in a spectrometer, the sampled volume is sufficiently large so that the system is optically homogeneous. Therefore, calibrations based on the average membrane potential of the cell population can be applied. The most common of these involves the use of a valinomycin-mediated K^+ diffusion potential which can be set by varying the potassium concentration in the external medium; the assumption here is that the potassium permeability will dominate that of any other ion in the presence of valinomycin, a potent and highly selective potassium ionophore. A variation of this approach seeks the "null point" at which the external K^+ is such that the addition of valinomycin causes no change in fluorescence; at the "null point", therefore, the equilibrium potential for potassium must equal the resting potential. Of course, if the relative ionic permeabilities and intracellular concentrations are known, direct calibration of the optical signal is straightforward; on the other hand, for cases where this condition is met, the dye methodologies are not really necessary, except as a convenient means to kinetically follow responses to external stimuli. The earlier review by Freedman and Laris[22] discusses calibration of cells in suspension in considerable detail; the chapter by J. C. Smith, which follows this chapter, discusses calibration of dyes applied to energy-transducing organelles.

For single cells or optically heterogeneous tissue preparations, microelectrode measurements are usually necessary to calibrate a dye. This always has the uncertainty that impalement of the cells can mechanically perturb the membrane potential. Also, the optical signal must originate from only the cell type of interest. This latter requirement effectively precludes calibration of absorbance changes, except for highly reproducible preparations or when the calibration can be performed on each preparation; this is because the optical signal contains variability from light scattering, sample opacity, or the level of dye staining. Fluorescence measurements are somewhat easier to calibrate with microelectrodes because the change in response to a voltage step can be normalized to the total resting fluorescence; thus, only light originating from the dye is measured and the relative nature of the measurement accounts for any variation in staining. Even with fluorescence, however, the presence of extraneous cells can produce variable background light signals which cannot be normalized away. This is the case, for example, with the squid axon preparation where Schwann cells contribute about 20 times as much fluorescence as the axon and the precise level of background depends on the dissection.[3]

An alternative for the calibration of membrane-staining dyes on cells with spherical or ellipsoidal shapes was described by Gross et al.[23] In this procedure, the membrane potential is induced with an external electric field and therefore does not require impalement of the cell. For simple cell geometries the membrane potential is readily calculated at a given point on the cell where the spectroscopic response is monitored.

Both the external field and microelectrode methods for membrane-bound dyes, only allow the calibration of membrane voltage *changes*. As discussed in Section II, the Nernstian dyes that we have recently developed[11,12] can be used to determine the potential on individual cells through a microscope. The principle behind these permeant cationic dyes is that the ratio of fluorescence intensities inside to outside the cells can be equated with the corresponding concentration ratio and substituted into Equation 1 to obtain the potential. In practice, two corrections must be applied to these measurements. The nonpotential-dependent binding of the dye to the membranes or organelles of the cell must be determined by measuring the fluorescence ratio of cells that are completely depolarized. This is accomplished by permeabilizing the cell or by clamping it to zero potential with high external potassium and valinomycin. The second kind of error which must be taken into account arises from the contribution of fluorescence from dye outside the cell to the fluorescence intensity measured when the focus is on the cell interior. This can be determined by measuring the fluorescence from a membrane-impermeant dye at this same focal plane and with the same measuring aperture. Once these corrections have been made, the Nernstian dyes require no special calibration since they are not self-quenched at the concentrations employed and their fluorescence efficiency is not significantly sensitive to environment.

IV. EVALUATION OF PROBE CHROMOPHORES

A. Cyanines

The defining features of a cyanine chromophore are depicted in Figure 1a. The positive charge characteristic of cyanines is completely delocalized and is equally shared by the two nitrogens. The chromophore is planar and has an additional plane of symmetry perpendicular to the molecular plane. The heterocyclic ring systems, which are typically indoles, benzoxazoles, or benzothiazoles, are linked by 3, 5, or 7 methines. This can provide a range of conjugations which allow these dyes to span the spectral range between 450 and 750 nm with extinction coefficients of about 10^5; emission maxima are generally 20 to 40 nm higher than the corresponding absorption wavelengths (a modest Stokes shift). The alkyl groups on the two nitrogens do not contribute to the spectral properties of these dyes, but, do play a crucial role in their interaction with cells or membrane vesicles. A convenient nomenclature which specifies the heterocycle, the number of methines, and the length of the alkyl chains was introduced by Waggoner and his co-workers,[4] and has been widely adopted:

$$di - X - C_m(n)$$

X = I, O, or S, corresponding to 3,3-dimethylindole, benzoxazole, or benzothiazole, respectively; C_m signifies the number, m, of carbons comprising the alkylchains; and n is the number of methines between the heterocycles.

One of the most widely used potentiometric cyanines, di-S-C_3(5), is depicted in Figure 1b.

The cyanines are slow dyes which adopt the redistribution mechanism in their response to membrane potential changes. The delocalized positive charge allows these probes to be permeable. The dyes do not collect inside the cells simply according to Equation 1, however, because the potential-dependent redistribution is invariably coupled to a variety of other equilibria; the most common of these include dye aggregation and dye association with the surface of the cell and internal membranes, but, interaction with specific proteins (e.g., hemoglobin in red cells) has also been noted. Care must be exercised in properly considering the effect of mitochondrial potential in applying these dyes. Dye aggregation is especially effective for the more conjugated benzothiazole members of this class of dyes (n = 5,7; X = S), at concentrations in the micromolar range. The cell and dye concentration must be optimized so that the potential-dependent accumulation leads to dye aggregation with fluorescence quenching. Binding of the dyes to the hydrophobic membrane environment leads to fluorescence enhancement; hyperpolarization leads to a fluorescence increase, therefore, for the indo- and oxacyanines (n = 3 and X = I, O) and when the membrane to dye ratio is high. The lengths of the alkyl chains appended to the nitrogens influence the strength of membrane binding (large m give stronger binding) and the rate of permeation (very long alkyl chains prevent the transbilayer flip of the probes). This provides another convenient handle for optimizing the performance of the cyanine probes in a particular application.

B. Merocyanines

These chromophores are characterized by an amino nitrogen conjugated to a carbonyl oxygen via a pi-system of variable length. A dipolar resonance structure in which the nitrogen is positively charged and the oxygen is negative (Figure 4a) is often an important contributor to the ground state electronic structure of merocyanines and may be the dominant structure in polar solvents. Platt[19] predicted that there may be a large change in the relative contributions of these two structures in the lowest energy-excited state. This has the consequence that these dyes should have spectra which are highly solvent-sensitive ("solvatochromism"); electrochromism is also a possibility.

The nitrogen on the positive end of the chromophore dipole can be incorporated into a variety of heterocyclic ring systems, including: indole, benzoxazole, and quinoline. The negative end of the chromophore has its oxygen attached, most commonly, to barbituric acid, isoxazolone, or rhodanine heterocycles. Various combinations of all these (as well as several others) have been used to construct voltage-sensitive dyes via appropriate appendage of alkyl chains or localized charges to anchor the probes to the hydrocarbon interior and the head-group region, respectively, of the membrane. This variety of merocyanine structures makes it impossible to devise a simple nomenclature; in practice, each dye has been identified by its catalog number (for commercial dyes) or by a code number assigned by the organic chemist who synthesized it (the systematic names of the dyes using the I.U.P.A.C. nomenclature rules are hopelessly complex).

The merocyanine probes are usually membrane stains with fast responses to potential changes. In the limited cases where mechanistic studies have been performed, reorientation appears to underlie the response.[2,17,18] The dipolar structure is ideally suited for reorientation if the chromophore can pivot on its hydrophilic anchor. This appears to be the case for M-540 (Figure 4b) where a benzoxazole is linked to a barbituric acid and a propylsulfonate serves as a pivot. Merocyanines also display a tendency to aggregate, presumably in an antiparallel array, and this may contribute to the response mechanism as well.

Merocyanines generally have high extinction coefficients and may be quite fluorescent. It is difficult to generalize about their molecular properties, however, because of the variety of structures which fit the definition and which have been tried as potentiometric probes in various applications. It does seem that the merocyanine probes generally appear problematic with respect to photodynamic effects and toxicity. They have, therefore, lost much of their initial popularity in fast dye applications. The chapters by Salzberg and Salama detail the utility of merocyanine-rhodanine and merocyanine-oxazolone probes in following voltage-dependent absorption changes.

C. Styryls

The *p*-aminostyryl-based probes (Figure 5) are currently the most popular fast-fluorescent probes. Most commonly, they consist of a quarternized pyridinium linked to an aniline via one or two double bonds. The chromophore bears a positive charge which is concentrated in the pyridine ring in the ground state and shifts to the aniline in the excited state.[6,9] This "charge shift" can couple with the electric field across the membrane if the probe is appropriately oriented, leading to an electrochromic mechanism. The chromophore can be oriented so that its long axis is perpendicular to the membrane axis via a pair of n-alkyl groups on the aniline nitrogen. The pyridinium nitrogen is a convenient location for an aqueous anchor, usually in the form of a propyl- or butyl-sulfonate. These fixed negative charges serve to prevent rapid transmembrane migration of the probe and also neutralize the positive charge of the chromophore so that the surface potential of the membrane is left largely unperturbed. The stability of these dyes, especially the less conjugated ones, is quite high and toxicity or photodynamic effects are low. We have tried to assign abbreviations to these dyes which are connected to their structures:

$$di - n - AANR$$

n represents the number of carbons in each of the alkyl chains; AA represents the amino aromatic (AS = aminostyryl; AB = anilinobutadienyl; ANE = aminonaphthethenyl); N represents the nitrogen heterocycle; (P = pyridinium; Q = quinolinium); R represents the group used to quarternize the nitrogen; (PS = propylsulfonate; BS = butylsulfonate).

The structure of di-6-ASPPS is presented in Figure 5b.

Most of the existing styryl dyes have absorption maxima within the range 450 to 550 nm when membrane-bound, with extinctions of about 4×10^4. The emission maxima are shifted 100 to 150 nm to longer wavelengths making these probes very convenient to use in highly scattering preparations. In addition, the fluorescence quantum yields are about 0.3 for the membrane-bound dyes and a factor of 100 to 1000 times less in water; this allows one to neglect the contribution from any unbound dye to the fluorescence (in any event, with n > 3 the binding to membranes is strong enough so that unbound dye can be minimized under most experimental conditions). The spectral and binding properties of many of these probes are fully detailed in Reference 10.

As noted above, these dyes were designed with electrochromism in mind and much evidence for this mechanism has been accumulated on a model-membrane system.[9,10] The largest relative fluorescence change in these experiments is 10%/100 mV for the probe di-4-ANEPPS. It is obtained by choosing excitation and emission wavelengths on the red edges of the spectral bands where the respective spectral responses reinforce each other. In addition to the ability to design probes with the assurance of a fast response, one of the original attractions of electrochromism was the expectation that the magnitude of the signal might be reasonably constant from preparation to preparation. This is because the chromophore interacts directly with the electric field in the membrane; other mechanisms require a delicate balance between various chemical states of the probe which can be tipped in one direction or the other by variations in the membrane environment, as well as the potential across it. Unfortunately, it is always possible that a response mechanism which is suppressed in one experiment, say a model membrane, may emerge in a different application of the same probe. This is the main reason, we believe, that our hopes for a universal electrochromic probe have proven unrealistic.[24] Still, di-4-ANEPPS has given remarkably consistent responses in a variety of systems and is currently among the most generally useful fast-fluorescent probes. Some dienyl probes have shown larger signals in particular experiments,[20] but these have been quite inconsistent. RH-421 (DI-5-ABPBS, by our nomenclature), for example, was shown to give responses in excess of 20%/100 mV by Grinvald et al.[20] in a neuroblastoma cell preparation, presumably because of a reinforcement, via some unknown mechanism, of the basal 5% response[10] associated with its electrochromism; this sensitivity is not conserved in other cultured excitable cells and is even quite variable for the neuroblastoma preparation.[26] (See also Reference 10 for an analysis of this issue in terms of probe structure.)

D. Oxonols

Oxonol chromphores contain two heterocyclic ring systems linked by a conjugated chain through which a negative charge is delocalized (Figure 2a). Originally incorporated into potentiometric membrane probes by Chance and his co-workers for studies of cells and organelles in suspension,[14] they have recently become popular for studies of individual excitable cells and neuronal networks as well (See the chapter by London et al.).

The most widely used oxonol dye for cells in suspension is OX-V which is depicted in Figure 2b. As detailed in the following chapter by Smith, OX-V and OX-VI (where the phenyls are replaced with propyl groups) have been especially successful in studies of energy-transducing organelles, but they are equally applicable to cells. Absorption maxima occur at about 610 nm with an extinction of 1.2×10^5; emission is at 645 nm. The negative charge is delocalized and therefore does not prevent membrane binding, but does make binding potential sensitive. The spectral properties are altered upon binding and most experiments have been performed by following absorbance changes. Aggregation does not appear to be important for these probes. The sensitivity can be maximized via the use of dual wavelength spectrophotometry, which also minimizes possible artifacts such as changes in turbidity. OX dyes can be used in fluorescence modes as well.

Oxonols are generally less cytotoxic than the cyanines. An oxonol used to stain a simple invertebrate neuronal network (London et al., Chapter 19), was also particularly photostable and showed virtually no photodynamic damage. These experiments are generally performed in absorption modes with quite large signals. The mechanism by which these absorption changes occur is not clear and deserves serious investigation so that improved membrane-staining oxonols may be designed.

E. Rhodamines

Rhodamines are among the oldest dye chromphores, containing a highly delocalized positive charge in a more compact structure (Figure 3) than the elongated dye molecules discussed above. They have absorption spectra with maxima in the range of 500 to 600 nm and extinctions of about 8×10^4; emission maxima are only about 20 nm higher than the excitation wavelengths and quantum yields are high, consistent with the rigid symmetrical structure of the chromophore (compare Chapter 1). The spectral properties are not very sensitive to environment and these dyes have very little tendency to aggregate.

The most popular member of this class of probes for potentiometric applications is Rhodamine 123 (Figure 3b) which is widely used to stain mitochondria in living cells.[25] This dye is relatively hydrophilic and therefore persists inside the highly negative mitochondria after the excess stain is washed away; it will be released slowly if the mitochondria are deliberately depolarized, but most applications have been aimed at simply visualizing the organelles under the microscope rather than monitoring their potential.

We have screened a number of positively charged rhodamines with the aim of taking advantage of their insensitivity to environment and low tendency to aggregate to develop a Nernstian fluorescent indicator (Section II.A). Rhodamine 123 was too impermeable to provide rapid reversible responses to potential changes. The other commercially available cationic rhodamines were too hydrophobic and had high background fluorescences due to membrane staining. We have therefore synthesized a pair of new rhodamines, the ethyl and methyl esters of tetramethylrhodamine, which provide the ideal combination of rapid membrane permeation and low membrane binding.[11,12] The key difference in our protocol for treating cells compared to that used for Rhodamine 123 staining of mitochondria, is that the dye is not washed away, but allowed to remain in equilibrium with the internal dye. Thus, the ratio of internal to external fluorescence intensities is a direct indicator of potential. Using the calibrations and corrections outlined in Section III.C, these probes allow the measurement of resting potentials on the plasma membrane of individual cells under the microscope.

REFERENCES

1. **Cohen, L. B., Salzberg, B., Davilla, H. V., Ross, W. N., Landowne, D., Waggoner, A. S., and Wang, C. H.**, Changes in axon fluorescence during activity: molecular probes of membrane potential, *J. Membr. Biol.*, 19, 1, 1974.
2. **Ross, W. N., Salzberg, B. M., Cohen, L. B., Grinvald, A. O., Davilla, H. V., Waggoner, A. S., and Wang, C.-H.**, Changes in absorption, fluorescence, dichroism, and birefringence in stained axons: optical measurement of membrane potential, *J. Membr. Biol.*, 33, 141, 1977.
3. **Gupta, R. K., Salzberg, B. M., Cohen, L. B., Grinvald, A., Lesher, S., Kamino, K., Boyle, M. B., Waggoner, A. S., and Wang, C. H.**, Improvements in optical methods for measuring rapid changes in membrane, *J. Memb. Biol.*, 58, 123, 1981.
4. **Sims, P. J., Waggoner, A. S., Wang, C.-H., and Hoffmann, J. F.**, Studies on the mechanism by which cyanine dyes measure membrane potential in red blood cells and phosphatidyl choline vesicles, *Biochemistry*, 13, 3315, 1974.
5. **Waggoner, A. S.**, Dye indicators of membrane potential, *Annu. Rev. Biophys. Bioeng.*, 8, 847, 1979.
6. **Loew, L. M., Bonneville, G. W., and Surow, J.**, Charge shift optical probes of membrane potential. Theory, *Biochemistry*, 17, 4065, 1978.
7. **Loew, L. M.**, Design and characterization of electrochromic membrane probes, *J. Biochem. Biophys. Meth.*, 6, 243, 1982.
8. **Hassner, A., Birnbaum, D., and Loew, L. M.**, Charge shift probes of membrane potential. Synthesis, *J. Org. Chem.*, 49, 2546, 1984.
9. **Loew, L. M. and Simpson, L.**, Charge shift probes of membrane potential. A probable electrochromic mechanism for ASP probes on a hemispherical lipid bilayer, *Biophys. J.*, 34, 353, 1981.
10. **Fluhler, E., Burnham, V. G., and Loew, L. M.**, Spectra, membrane binding and potentiometric responses of new charge shift probes, *Biochemistry*, 24, 5749, 1985.
11. **Loew, L. M., Ehrenberg, B., Fluhler, E., Wei, M.-d., and Burnham, V.**, A search for nernstian dyes to measure membrane potential in individual cells, *Biophys. J.*, 49, 308a, 1986.
12. **Ehrenberg, B., Wei, M.-D., and Loew, L. M.**, Nernstian dye distribution reports membrane potential in individual cells, in *Membrane Proteins*, Goheen, S. C., Ed., Bio-Rad Laboratories, Richmond, Calif., 1987, 279.
13. **Waggoner, A. S. and Grinvald, A.**, Mechanisms of rapid optical changes of potential sensitive dyes, *Ann. N.Y. Acad. Sci.*, 303, 217, 1977.
14. **Smith, J. C., Russ, P., Cooperman, B. S., and Chance, B.**, Synthesis structure determination, spectral properties, and energy-linked spectral responses of the extrinsic probe oxonol V in membranes, *Biochemistry*, 15, 5094, 1976.
15. **Loew, L. M., Benson, L., Lazarovici, P., and Rosenberg, I.**, A fluorometric analysis of transferrable membrane pores, *Biochemistry*, 24, 2101, 1985.
16. **Lichtstein, D., Kaback, H. R., and Blume, A. J.**, Use of lipophilic cation for determination of membrane potential in neuroblastoma hybrid cell suspensions, *Proc. Natl. Acad. Sci. U.S.A.*, 76, 650, 1979.
17. **Dragsten, P. R. and Webb, W. W.**, Mechanism of the membrane potential sensitivity of the fluorescent membrane probe merocyanine 540, *Biochemistry*, 17, 5228, 1978.
18. **Verkman, A. S. and Frosch, M. P.**, Temperature-jump studies of merocyanine 540 relaxation kinetics in lipid bilayer membranes, *Biochemistry*, 24, 7117, 1985.
19. **Platt, J. R.**, Electrochronism, a possible change of color producible in dyes by an electric field, *J. Chem. Phys.*, 25, 80, 1956.
20. **Grinvald, A., Fine, A., Farber, I. C., and Hildesheim, R.**, Fluorescence monitoring of electrical responses from small neurons and their processes, *Biophys. J.*, 42, 195, 1983.
21. **Lojewska, Z. and Loew, L. M.**, Insertion of amphiphilic molecules into membranes is catalyzed by a high molecular weight non-ionic surfactant, *Biophys. Biochim. Acta*, 899, 104, 1987.
22. **Freedman, J. C. and Laris, P. C.**, Electrophysiology of cells and organelles: studies with optical potentiometric indicators. Membrane research: classic origins and current concepts, *Int. Rev. Cytol. Supp.*, 12, 177, 1981.
23. **Gross, D., Loew, L. M., and Webb, W. W.**, Optical imaging of cell membrane potential: changes induced by applied electric fields, *Biophys. J.*, 50, 339, 1986.
24. **Loew, L. M., Cohen, L. B., Salzberg, B. M., Obaid, A. L., and Bezanilla, F.**, Charge shift probes of membrane potential. Characterization of aminostyrylpyridinium dyes on the squid giant axon, *Biophys. J.*, 47, 71, 1985.
25. **Johnson, L. V., Walsh, M. L., Bockus, B. J., and Chen, L. B.**, Monitoring of relative mitochondrial membrane potential in living cells by fluorescence microscopy, *J. Cell Biol.*, 88, 526, 1981.
26. **Pine, J.**, personal communication.

Chapter 15

POTENTIAL-SENSITIVE MOLECULAR PROBES IN ENERGY-TRANSDUCING ORGANELLES

J. C. Smith

TABLE OF CONTENTS

I. INTRODUCTION

This chapter deals with the application of potential-sensitive optical probes to the broadly defined discipline of bioenergetics. Although in principle any system containing an ATPase can be regarded as falling into the latter category, the material contained herein is largely confined to the application of probes to energy transducing organelles. The behavior of optical molecular probes in model systems and in organ or tissue-level studies is cited only when the model system studies directly bear on the behavior of the probes in these organelles or in the latter work when the optical signals appear to be primarily due to such organelles. The use of optical indicators in model systems and in organ or tissue-level studies is covered elsewhere in this work. Whole-cell probe applications are covered by Freedman and Laris in Volume 3, Chapter 16. The contents of this chapter are based on literature searches through March 1986.

Most of the references and published information are organized in the several tables. References have been grouped into the following categories: mitochondria and related preparations, Table 1; photosynthetic systems, Table 2; sarcoplasmic reticulum, Table 3; and several miscellaneous studies, Table 4. In each table, the author(s), probe(s) employed, and a reference number are provided. The body of the text contains a discussion of several current issues regarding the application of potential-sensitive molecular probes: the specificity or lack thereof of the dye to membrane potential(s), the mechanisms by which the probes respond to charge separation, the nature of the charge separation to which the probes are sensitive, calibration procedures and attendant problems, and the kinetic competence of selected probes to follow the charge-separation events in membranes. Although an effort has been made to provide a comprehensive listing of papers involving probe usage, those studies in which the behavior of the probe in the membrane system of interest is critically examined have been given priority in the accompanying text. The application of extrinsic probes to a variety of biological preparations, including energy-transducing membranes, has been considered in reviews by Waggoner,[1] Cohen and Salzberg,[2] Bashford and Smith,[3] and Bashford.[4] Early application of a number of oxonol probes in energy-transducing membrane systems has been described in monograph articles by Chance et al.,[5] Bashford et al.,[6] and Smith et al.[7] The utility of a number of commonly employed experimental techniques for measuring membrane potentials in mitochondria, including molecular probes, has been reviewed by Tedeschi.[8]

The probes are usually charged and often consist of fused ring structures, such as those characteristic of ANS, or the berberine series, or fall into the polyene group in which two ring structures of varying complexity are connected by a conjugated carbon chain. In the latter group, the wavelength of maximum absorption is a function of the length of the conjugated chain[9] which can be varied to obtain dye derivatives, the optical absorption and emission spectra of which do not significantly overlap those of intrinsic membrane pigments. With the exception of carotenoids such as β-carotene and certain intrinsic energy-dependent probes such as the cytochrome c oxidase,[10] these probes must be added to the system of interest in a carefully controlled fashion in order to obtain maximum energy-dependent optical signals with minimal effects on membrane viability.

II. ENERGY-TRANSDUCTION MODELS

Because the interpretation of molecular probe optical signals in response to light or substrate activation of energy-transducing preparations in terms of mechanistic and kinetic schemes involves, to some extent, the chosen model for the energy-transduction process, and, conversely, the behavior of the probe signals under energy-transducing conditions can provide insight into the validity of such models, this section presents a brief summary of several current proposals concerning energy transduction that are applicable to the areas covered in this chapter. The most commonly accepted model for energy transduction in mitochondria and preparations derived therefrom, as well as photosynthetic systems and a host of single-cell organisms is the chemiosmotic theory of Mitchell,[11-13] in its original form or as modified by the introduction of Q cycles[14] and variable $H^+/\partial e^-$ stoichiometries. This theory is based on the presence of a transmembrane electrical potential difference $\Delta\Psi$ and a pH-gradient ΔpH related according to Equation 1 generated by vectorial electron transport and the extrusion of protons to the bathing medium or proton uptake by the thylakoid space resulting in the mitochondrial matrix or the latter space of chloroplasts, respectively, being at a negative or positive potential relative to that of the external phase:

$$\Delta\mu_H = \Delta\Psi - 2\cdot 3\,\frac{RT}{F}\,\Delta pH \tag{1}$$

Based on recently obtained data on the lateral diffusion of enzymes in energy-transducing membranes, Boyer[15] and Slater et al.[16] have suggested a direct interaction of energized redox enzymes with the ATP synthase through a transient, i.e., collisional, process in which energy is transferred to the ATPase, driving a conformational change in the latter complex that results in the spontaneous synthesis of ATP from bound ADP and P_i and the release of previously formed ATP from the tight binding site. In the model of Slater et al., the energized redox enzyme, the ubiquinol cytochrome c oxidoreductase in mitochondria, can either relax to the unergized enzyme form acting as a proton pump in the relaxation process that produces a $\Delta\mu_H$ or drive a conformational change in the ATP synthase that may then either directly synthesize ATP as described above or function as a proton pump producing $\Delta\mu_H$.

More recently, however, the role of surface potentials in mitochondrial oxidative phosphorylation has received considerable attention. Malpress[17] has proposed that electron transport results in the formation of fixed negative charges on or near the mitochondrial inner membrane surface. The resulting surface potential controls the concentration of ions in the diffuse double layer; the electrochemical gradient or proton motive force in this model is related to the surface electrostatic potential $\Delta\Psi_H$ by a factor f which is a function of the prevailing experimental conditions and substrates used in oxidative phosphorylation:

$$\Delta\mu_H/F = f\Delta\Psi_H \tag{2}$$

$\Delta\mu_H$ is derived from proton movements into the bulk phase from the double-layer region which are the net effect of all competing ion movements which affect the proton concentration level in the double layer. The latter sensitivity has been suggested as an explanation for certain observations that are not in accord with the chemiosmotic model, such as the dependence of the ratio of the phosphorylation potential to the electrochemical gradient $\Delta\mu_H$ on the value of the electrochemical gradient itself. Woelders et al.[18] have, however, presented evidence that such dependence is due to an experimental artifact; when the later artifact is eliminated $\Delta G_p:\Delta\mu_H$ is constant over a twofold range of $\Delta\mu_H$. Kell[19] has proposed a somewhat similar model in which the existence of a transmembrane potential that exceeds the potetial difference between the matrix and bulk aqueous phases is suggested. Most of the functional proton translocation is supposed to occur at the surface of the mitochondrial membrane in fixed channels which contain the ATPase at a potential different from that of the bulk phase. Proposed mechanisms include a Grotthus transfer of protons between adsorbed water molecules. Like the Malpress model, the electrodic model of Kell envisions the formation of surface charges presumably due to changes in ionization, pK, or conformation of proton carriers being linked to electron transport. Van Walraven et al.[20] have suggested that the effective charge gradient is either within the bilayer dielectric or across the diffuse double-layer region with the potential difference between the two bulk phases serving as a "buffer" to drive ion translocation, or to smooth out fluctuations in ATP synthesis or electron flow. See, however, the work of Robertson and Rottenberg,[21] of Matsuura et al.,[22] and of Masamoto et al.[23]

III. MITOCHONDRIA AND RELATED PREPARATIONS

Among the earliest applications of polyene type potential-sensitive probes to mitochondria is the work of Laris et al.[24] in which the cyanine diS-C_3-(5) was employed in hamster liver preparations. A nominal 10 % loss in fluorescence yield relative to the control level was observed upon addition of succinate or ATP; the succinate-dependent optical signal was virtually completely reversible by cyanide, valinomycin, or dinitrophenol and the ATP-dependent one by oligomycin or dinitrophenol. The diS-C_3-(5) fluorescence yield was a linear function of valinomycin-generated diffusion potentials in which the external K^+ concentration was in the range of 1 to 40 mM. The dye response, however, became nonlinear at concentrations below 1 mM and appeared to tend toward saturation at lower K^+ concentrations. Using a value of 100 mM for the internal K^+ concentration, the same as that used by Harris and Pressman[25] for rat liver mitochondria, a membrane potential of -64 mV was found for the resting state of mitochondria in a K^+ medium and -45 mV for a suspension in Na^+ medium; similar values have been reported by Akerman and Wikström[26] for the uncoupled state of this preparation (see below). A value of -150 to -180 mV was determined using the same internal concentration of K^+ when succinate was the substrate.

Zinchenko et al.[27] have reported a similar value for $\Delta\Psi$ in mitochondria undergoing oxidative phosphorylation based on fluorescence measurements employing the probes diS-C_2-(5) and diS-C_3-(5) which were again calibrated with K^+ diffusion potentials. These authors reported only a small decrease (3 to 8 mV) in $\Delta\Psi$ during ATP synthesis, but, a drop of 10 to 150 mV during Ca^{++} transport.

The mechanism of diS-C_3-(5) absorption spectrum and fluorescence signals generated in pigeon heart mitochnondria (PHM) by succinate, glutamate, or ATP-Mg^{++} consumption have been investigated by Bammel et al.[28] Detailed equilibrium-binding studies using the methods of Bashford and Smith[29] have indicated a substantial reduction in the number of binding sites available to the dye when substrate is present relative to that available when the mitochondria are inhibited by cyanide. The binding-site reduction apparently reflects alterations in the membrane structure at the level of the probe-bonding sites, and may be related to membrane structural alterations originally described by Packer[30] and Packer and Tappel[31] using light-scattering techniques. Dis-C_3-(5) also inhibited both the signal amplitude and rate at which the ATP-dependent cytochrome c oxidase soret-band shift developed and if sufficient dye was added, this signal could be completely abolished. These observations indicate that the probe moves across the mitochondrial inner membrane by an electrophoretic process that competes with the process responsible for the soret-band shift for energy derived from ATP hydrolysis and ultimately accumulates in the matrix volume, a process that is consistent with dye ejection from the membrane suggested by the binding studies previously cited.

Using rapid-mixing techniques, the time course of the glutamate-driven diS-C_3-(5) absorption spectrum change has been monitored in PHM.[28] The time course of the glutamate-dependent signal can be fit to a single exponential function with a typical halftime of 1 sec; at fixed-dye concentrations, the rate constant characterizing this dye signal was independent of membrane concentration; at fixed-membrane concentration, however, the rate constant was inversely dependent on dye concentration in the range of 2 to 120 nmol dye per mg PHM protein. The later behavior appears to be related to the energy-dependent inhibition of NADH dehydrogenase activity reported by Connover and Schneider[32] for the ethyl derivative of diS-C$_3$-(5). Measurements in the author's laboratory using diS-C$_3$-(5) itself indicate that glutamate-driven oxygen consumption in the PHM suspension is inhibited by the dye, but no effect on the uncoupled-state respiration rate was observed. Pechatnikov et al.[33] have also reported rotenone-like inhibition of respiration in the same preparation by both diS-C$_2$-(5) and diS-C$_3$-(5). Montecucco et al.[34] have also found that a number of cyanine probes, including: diS-C$_3$-(5), diS-C$_2$-(5), diO-C$_2$-(3), and diI-C$_6$-(5) were inhibitors of Site I in rat liver mitochondria and that the inhibition in the case of diS-C$_3$-(5), which was studied in detail, was likely a consequence of the accumulation of this probe in the matrix volume of the organelle, which accumulation in turn was dependent on the presence of a membrane potential. None of the probes investigated inhibited Site II. Aiuchi et al.[35] have reported respiratory inhibition by diS-C$_3$-(5) in synaptosomes.

diS-C$_3$-(5) has also been employed by Quintanilha and Packer[36] as a presumed qualitative indicator of the transmembrane potential generated by ATP hydrolysis in mitoplasts, a mitochondria preparation in which the outer membrane has been removed. The distribution of the cationic spin label CAT$_{12}$ between the aqueous and membrane phases was used by these investigators to determine surface charge density changes in the inhibited and ATP-energized preparation; these results suggest that additional negative charges were developed on the outer surface of the mitoplast that rendered this surface 20 mV more negative in a sucrose medium and 17 mV more negative in a KCl medium when ATP was present than when the mitoplasts were inhibited by oligomycin. Surface potential changes were suggested as a means of controlling the electrochemical gradient across the inner mitochondrial membrane, as also proposed by Malpress.[17] Additional work with intact mitochondria from Vitamin E-deficient animals indicate that the surface potential was slightly more negative than that in control animals.[37]

Bakeeva et al.[38] have found that the respiration rate in intact lymphocytes treated with oligomycin is increased by an order of magnitude by DNP at 40 to 60 μM concentration, as well as by palmitate and oleate, but, not stearate at concentrations 3 to 4 times that of DNP, indicating that the mitochondria in these cells maintain a high electrochemical gradient. The presence of bovine serum albumin was required for the fatty acid stimulation of respiration. In mitochondria isolated from these cells, the onset of the uncoupling effect of DNP and of the two fatty acids occurred at lower concentrations than observed in intact cells, provided that the mitochondria-isolation media contained bovine serum albumin. An increase in the diS-C$_3$-(5) fluorescence yield in intact lymphocytes occurred at DNP concentrations as low as 2 to 4 μM and was maximal at uncoupler concentrations near 40 μM. Due to interference with the probe fluorescence, analogous studies with palmitate could not be performed.

Michejda et al.[39] and Adamski et al.[40] have employed diS-C$_3$-(5) as a qualitative probe of $\Delta\Psi$ in mitochondria from *Acanthamoeba castellani*. Succinate or NADH induced a fluorescence decrease that was reversed by valinomycin + K$^+$ or FCCP. ATP, however, indued a significantly smaller dye signal than the former substrates, apparently due to impared ATP/ADP transport due to low endogenous pools of adenine nucleotides in this preparation. The decrease in diS-C$_3$-(5) fluorescence yield induced by malate was only partially reversed by cyanide and again decreased by subsequent addition of AMP and finally completely abolished by salicylhydroxamate (SHAM) or rotenone. The dye response to stimulation of the alternate pathway by AMP could only be observed with malate, as substrate indicating that this pathway does not generate a membrane potential except at Site I, as also suggested by Moore[41] based on investigations in the mung bean mitochondria system. In contrast to the observations of Connover and Schneider[32] and Bammel et al.,[28] diS-C$_3$-(5) did not appear to inhibit the mitochondrial NADH dehydrogenase in this preparation under the prevailing experimental conditions.

Laris[42] has observed a transient decrease in diS-C$_3$-(5) fluorescence yield in mitochondria treated with rotenone and oligomycin when ATP was added. Atractyloside, bongkrekic acid, and ADP all blocked the transient response of the probe to ATP addition that appears to generate an inside-negative potential. Addition of ADP after the addition of ATP leads to a transient increase in dye fluorescence yield, indicating the presence of a depolarizing potential that was again sensitive to atractyloside and bongkrekic acid. These observations indiate that the ATP/ADP exchange process is electrogenic in mitochondria. A similar series of investigations has been carried out using diS-C$_3$-(5) in inhibited submitochondrial particles with an intact F$_1$ by Lauquin et al.[43] that suggest an electrogenic adenine nucleotide exchange mechanism in this preparation as well.

Akerman and Wikström[26] have employed safranine, a cationic cyanine probe, in investigations of energy transduction in rat liver mitochondria. These investigators have demonstrated that the probe absorption spectrum shift signal generated by succinate consumption is increased by nigericin and subsequently returned to the control level by valinomycin indicating that this probe is specifically sensitive to the membrane potential portion of the electrochemical gradient. The potential-dependent safranine optical signal is a linear function of the valinomycin-induced K$^+$ diffusion potetial values in the range 50 to 170 mV on the assumption that the K$^+$ distribution obeys the Nernst relationship. In agreement with Nicholls,[44] a potential of approximately 50 mV has been detected in uncoupled mitochondria which is presumably canceled by a pH gradient of the same magnitude, but, of opposite polarity. A value of 165 ($\pm$ 5) mV is reported for the membrane potential generated by succinate oxidation in the absence of P$_i$, based on the calibration of the safranine absorption spectrum shift by diffusion potentials as previously described. The value of the membrane potential supported by ATP hydrolysis was dependent on the magnitude of the phosphate potential but not on the ATP concentration itself.

Akerman[45] has also employed safranine to monitor the effect of Ca^{++} uptake in rat liver mitochondria; Murexide® or Arsenazo® III was used to measure the free Ca^{++} concentration. A decrease in the mitochondrial membrane potential is observed upon Ca^{++} uptake, which is eventually followed by a slow increase in the membrane potential, the rate of which can be stimulated by the addition of EGTA or phosphate. The initial rate of Ca^{++} uptake was a linear function of the membrane potential. This rate and the corresponding decrease in membrane potential were half maximal at 10 μM Ca^{++} and exhibited saturation at or above 20 μM Ca^{++}. Ca^{++} diffusion potentials generated by the addition of chelators when compared to the corresponding valinomycin-K$^+$ potentials suggest that calcium is transported with one net negative charge under the prevailing experimental conditions.

In a detailed study of the behavior of safranine in intact mitochondria, Zanotti and Azzone[46] have suggested that, in the presence of a membrane potential, this probe permeates the inner mitochondrial membrane and stacks on the matrix side of this membrane. The magnitude of the potential-dependent absorption spectrum shift signal was found to be a function of the dye to membrane protein concentration ratio; the probe response tended to become nonlinear at high valinomycin-generated K$^+$ diffusion potential values and the value of the potential at which the onset of the nonlinear probe response was observed was also a function of this ratio. High probe to membrane concentration ratios tended to produce both larger potential-dependent optical signals and to extend the potential range over which the probe response was linear. The half time of the development of the safranine signal fell in the 10 to 40 sec range and increased as the probe to membrane protein concentration ratio decreased. Thus, the conditions for the fastest probe response time to charge separation and those for the maximum potential value range over which the probe response was linear tended to be mutually exclusive in the mitochondria preparation. At fixed safranine to membrane protein concentration ratios, the rate at which the probe response developed increased linearly with the value of the membrane potential in the range of 50 to approximately 170 mV, as expected for an electrophoretically driven dye permeation of the mitochondrial membrane. This paper also contains a detailed examination of the conditions under which safranine is toxic and the nature of the toxicity in mitochondria.

Wilson[47] has investigated the behavior of the cyanide-insensitive electron transport system in *Phaseolus aureus* L. (mung-bean) mitochondria, using both safranine and ANS as extrinsic probes of membrane potential. Quinone substrates were employed that specifically exclude the involvement of phosphorylation at Site I, either directly or via Site-I-linked substrates produced by the primary oxidation process. The mitochondria were also treated with oligomycin to preclude the use of endogenous ATP. Under these conditions, ATP was produced as detected by a luciferase assay; both of the probes also qualitatively indicated the presence of a SHAM-sensitive $\Delta\Psi$ component at cyanide concentrations sufficient to completely inhibit cytochrome c oxidase. It was thus concluded that the cyanide-insensitive electron transport system is coupled to ATP production; this conclusion differs from that of a number of other investigators, notably Moore.[41] The source of these differences is discussed.

Kauppinen and Hassinen[48] have employed the safranine probe in studies of the Langendorff perfused rat heart. Double-wavelength reflectance spectroscopy could not be used in these measurements because the probe signal was masked by the high concentration of endogenous pigments in heart tissue, primarily myoglobin. Safranine fluorescence corrected for reflectance and motional artifacts was shown to be a sensitive and selective qualitative indicator of the mitochondrial membrane potential in this preparation. This conclusion was demonstrated by the probe fluorescence response to alterations in the extracellular K^+ concentration. Increasing this concentration to 18 mM decreases the plasma membrane potential, but, also increases the mitochondrial potential gradient due to a decrease in ATP consumption caused by stopping the mechanical activity of the heart. Under these conditions, a characteristic quenching of the probe fluorescence yield on the time scale of electron transport in the respiratory chain was observed which is indicative of mitochondrial membrane hyperpolarization. Furthermore, when the heart was maintained quiescent by Ca^{++} omission from the perfusing medium, alterations in the K^+ concentration did not affect the safranine fluorescence yield, as would be expected if the plasma membrane potential were contributing to the probe signal. Addition of the uncoupler CCCP to the perfusate caused an irreversible increase in the probe fluorescence yield. The readmission of Ca^{++} into the perfusate after a period of Ca^{++}-free perfusion of the isolated heart caused a very rapid collapse of the mitochondrial membrane potential, which could be readily followed by a pronounced increase in the safranine fluorescence yield. The direction of the probe changes was qualitatively similar for membrane potential changes calculated from the distribution of the permeable cation $TPMP^+$, indicating that a quantitative correlation between the probe fluorescence yield and the membrane potential of *in situ* mitochondria probably exists.

Riley and Pfeiffer[49] have recently reported that when rat liver mitochondria oxidizing succinate are allowed to accumulate Ca^{++} in excess of 40 nmol/mg protein and are then treated with excess EGTA, a fraction of the accumulated Ca^{++}, approximately 1 nmol/sec/mg protein, is rapidly released by a proposed reverse uniport process that results in an increase in the mitochondrial membrane potential that was monitored by safranine by means of dual-wavelength absorption spectroscopy. The passive back diffusion of protons is suggested as a means of eventual charge compensation. The rapid-release process is accompanied by an inhibition of O_2 consumption that is explained on the basis of the excessive energy requirement for proton extrusion from the mitochondria against the membrane potential, an obligatory process for respiration to occur.

In rat liver mitochondria treated with rotenone, Cockrell[50] has demonstrated ruthenium red-insensitive Ca^{++} uptake, driven by an artificial pH gradient generated by H^+/K^+ exchange, using the electroneutral ionophore dianemycin. Ca^{++} fluxes were measured using either antipyrylazo III or arsenazo III via absorption spectroscopy; proton fluxes were monitored either by a glass electrode or using phenol red, and the membrane potential was measured with safranine separately calibrated with valinomycin-generated K^+ diffusion potentials. Dianemycin-mediated Ca^{++} uptake is accompanied by an increase in $\Delta\Psi$, as monitored by safranine absorbance at 502 to 524 nm. The increase in $\Delta\Psi$, however, was not dependent on the presence of exogenous Ca^{++} since the same potential-sensitive probe response was observed when dianemyhcin was added in the absence of Ca^{++}. The safranine signal could be reversed by approximately 70% using FCCP. The increase in $\Delta\Psi$ was apparently derived from residual respiration that could be detected by O_2 uptake. The Ca^{++} uptake driven by the exchange process was inhibited by the local anesthetic nupercaine, by NH_3, P_i, and acetate. These observations suggest that the pathway for Ca^{++} influx may be the reverse of that responsible for ruthenium red-insensitive Ca^{++} release in this preparation.

Kim et al.[51] have investigated Ca^{++} transport *in situ* using digitonin-permeabilized cells of the protozoan *Tetrahymena pyriformis* GL. In the presence of oxidizable substrates and P_i, mitochondria were able to accumulate large amounts of Ca^{++} (3.5 to 5 μmol/mg of mitochondrial protein), without subsequent uncoupling and mitochondrial damage. The uptake of Ca^{++} was dramatically inhibited by micromolar quantities of the calcium indicator chlortetracyclin, which, under aerobic conditions caused uncoupling of respiration in Ca^{++}-loaded mitochondria. Under hypoxic conditions, in which the delivery of O_2 from the air to the stirred incubation medium became rate-limiting, oscillations in Ca^{++} uptake and in the mitochondrial membrane potential, monitored by diS-C_3-(5) fluorescence, were observed. Under the latter conditions, chlortetracyclin uncoupling of respiration and the Ca^{++} transport oscillations were prevented by ruthenium red and by EGTA. Two Ca^{++} transport pathways, a uniport system, and a chlortetracyclin-mediated electroneutral exchange process are suggested as an explanation of Ca^{++} transport oscillations under conditions of limited respiration. A number of qualitative differences in this oscillatory phenomenon and those observed in mammalian mitochondria studies, in which Ca^{++} chelators or large mitochondrial matrix volume changes were involved, are noted.

The probe DSMP has been shown in early work by Bereiter-Hahn[52] to respond to energization of PHM by various substrates or by ATP with an inhibitor-sensitive 8.5-fold increase in fluorescence yield that was interpreted as resulting from an increase in the probe quantum yield, due to the translocation of the dye to a more hydrophobic site, with no change in the amount of bound dye per unit weight of mitochondrial protein being observed upon energization of the preparation. The accumulation of DSMP within the mitochondria of cultured *Xenopus laevis* tadpole heart cells could be detected by fluorescence microscopy. Relative to the fluorescence signal from cell suspensions in succinate-supplemented media, pentachlorophenol caused a marked reduction in the corrected fluorescence yield of this probe. A corresponding cyanide-induced reduction in the fluorescence signal could not be demonstrated, possibly because of a direct effect of this inhibitor on the probe.

In more recent work with DSMP, Mewes and Rafael[53] have shown that the ^{3}H-labeled probe does distribute across the inner membrane of rat liver mitochondria, the fluorescence signal being linear with the amount of dye taken up by the organelle, up to 2 to 3 nmol of bound DSMP. The kinetics of the fluorescence-yield increase followed apparent first-order kinetics at fixed mitochondrial protein concentration over the rather narrow range of 0.73 to 1.16 μM. In agreement with Bereiter-Hahn,[52] the energy-dependent DSMP fluorescence-yield increase was ascribed to the transfer of the probe to a more apolar environment. Based on the assumption that the probe distribution obeys the Nernst relationship, a value of 190 mV, uncorrected for activity factors, was obtained.

Rafael and Nicholls[54] have demonstrated, using fluorescence microscopy, that DSMP is accumulated in the mitochondria of adipocytes from the cold-adapted guinea pig. The accumulation of the ^{3}H-labeled probe in the cell interior was decreased by valinomycin + K^+ and by rotenone and enhanced by nigericin, indicating that the distribution of this probe contained a component due to the presence of a mitochondrial potential. A 190 mV potential was estimated from the probe distribution, which in this case is governed by the sum of the plasma and mitochondrial membrane potentials. This value is again uncorrected for activity factors. Noradrenaline was shown to decrease the uptake of DSMP by an amount corresponding to a 15 mV decrease in the sum of the plasma and mitochondrial membrane potentials. Since a drop in $\Delta\Psi$, of 20 mV is sufficient for the transition to uncontrolled respiration in isolated brown adipose tissue mitochondria, the observed 15 mV decrease, if primarily due to a reduction in the mitochondrial potential, could explain the noradrenaline-induced stimulation of the respiration rate observed with the adipocytes. The increased respiration rate was suggested as arising from a proton conductance mediated by an uncoupling protein, since an additional decrease in the uptake of DSMP was observed when both noradrenaline and oligomycin were employed in combination, thereby essentially eliminating the possibility that the proton conductance is located in the ATP synthetase.

Using fluorescence microscopy, Korchak et al.[55] have demonstrated that the diO-C_6-(3) fluorescence yield loss generated by f-met-leu-phe and phorbol myristate acetate in human neutrophils originates primarily from the mitochondria and that the loss can be largely eliminated by pretreating the preparation with the electron transport inhibitors cyanide and rotenone or with the uncouplers CCCP and dinitrophenol; oligomycin, however, was without effect on the fluorescence yield loss. The difference in membrane potential results obtained using diO-C_6-(3) and the lipophilic cation TPMP$^+$ was explained on the basis of differences in probe distribution.

Johnson et al.[56,57] have demonstrated that a number of cationic dyes, including diO-C_2-(3) and related cyanines, safranine O, and several rhodamines are preferentially accumulated in the mitochondria of culture-grown cells of several types, whereas several neutral or anionic probes did not exhibit mitochondria-specific fluorescence; see Table 1 for a list of the probes employed in this work that specifically stained the mitochondria. The enhanced rhodamine 123 fluorescence yield associated with the stained *in situ* mitochondria, which was studied in detail, could be diminished by the uncouplers CCCP and FCCP, and by electron transport inhibitors such as cyanide, azide, antimycin A, and rotenone. A substantial increase in the fluorescence yield was induced by the introduction of nigericin to the cells under observation. The mitochondria in cells grown under anaerobic conditions were only weakly stained, but the fluorescence yield reached that associated with aerobically grown cells when oxygen was supplied. Oligomycin, however, did not affect the fluorescence yield of cells prestained with rhodamine 123; relative to the prestained fluorescence yield level, however, a slight increase in the dye fluorescence was observed when this ATPase inhibitor was introduced prior to staining with the rhodamine probe. These observations suggest that the accumulation of the cationic probes is govered by the *in situ* mitochondrial membrane potential. The degree of homogeneity in the staining of the mitochondria varied significantly among the several types of cells employed in this study, suggesting in the case of heterogeneous staining that pools of mitochondria at different membrane potentials may be present *in situ*. In related work, Davis et al.[58] have compared the uptake of rhodamine 123 and TPP in transformed MCF-7 cells from human breast adenocarcinoma with that in CV-1 cells from the African green monkey kidney epithelium. In media containing either 137 mM or 3.6 mM K$^+$ uptake and retention of the rhodamine probe, monitored by fluorescence microscopy, was elevated in the MCF-7 cells relative to that in the normal CV-1 cell line, suggesting increased plasma and/or mitochondrial membrane potentials in the former cell line. The difference in rhodamine and TPP uptake in the two lines at the higher K$^+$ concentration that dissipates the plasma membrane potential indicates an elevated mitochondrial $\Delta\Psi$ in the MCF-7 cells. Neither the TPP nor the rhodamine uptake was increased by nigericin in the presence of oubain, used to avoid a plasma membrane potential increase, in the MCF-7 line, but the uptake was markedly increased in the CV-1 line, suggesting that $\Delta\Psi$ is the primary contributor to the mitochondrial $\Delta\mu_H$ in the former cell type, whereas ΔpH may be appreciable in the CV-1 line. When nigericin alone was was used in the CV-1 line, both the plasma and mitochondrial membrane potentials were maximal and the uptake of TPP approached that observed in the MCF-7 line. From the ratio of TPP uptake at the high and low K^+ concentrations, it was concluded that the plasma membrane potential was also higher in the MCF-7 cells than in the CV-1 line, and that mitochondrial staining in the MCV-7 line by the rhodamine dye was enhanced at the lower K^+ concentration because of a preconcentration of the probe in the cell cytoplasm. Since rhodamine 123 appears to be selectively toxic to carcinoma cells, the use of this probe as a therapeutic agent is discussed.

Using the fluorescence microscopy technique, Divo et al.[59] have demonstrated that rhodamine 123 is taken up by the single mitochondrion present in the malarial parasite *Plasmodium falciparam* over the entire intraerythrocyte cycle of the parasite, indicating that the organism maintains a high mitochondrial membrane potential over this cycle. A portion of this work is illustrated in the photomicrographs in Figure 1. The retention of rhodamine 123 was reversed by the uncouplers DNP and CCCP, by valinomycin and nigericin, by the ATPase inhibitors DCCD, oligomycin, and quercetin, as well as by electron transport inhibitors antimycin A, HQNO, rotenone, azide, and cyanide. Treatment of the stained parasites with uncouplers reduced the probe fluorescence signal from both the cytoplasm and the mitochondrion, as expected for the dissipation of both the plasma and mitochondrial membrane potentials. Electron transport inhibitors, as expected, diminished only the mitochondrial-associated probe fluorescence yield, whereas DCCD at low concentration caused the organelle fluorescence yield to increase, since the ATP synthase activity is blocked and $\Delta\Psi$ is allowed to increase. The possible physiological role of the mitochondrial membrane potential is discussed in the context of the finding from previous investigations that glycolysis is the principal source of ATP in this parasite.

Kovac and Varecka[60] have employed diS-C$_3$-(5) in qualitative studies of membrane potentials in the yeast *Saccharomyces pombe;* both wild-type cells and respiration-deficient mutants were employed. The addition of the pore-forming agent nystatin, electron transport inhibitors azide or antimycin A, ATPase inhibitors DCCD and oligomycin, and the uncoupler CCCP to the wild-type cells at equilibrium with the optical probe caused a marked increase in the dye fluorescence yield, indicating the dissipation of either plasma or mitochondrial membrane potentials, or both. The pronounced effect of the specific electron transport inhibitors indicates that the composite membrane potential to which the probe is responding has a major mitochondrial potential component. D-glucose and other hexokinase substrates induced a biphasic diS-C$_3$-(5) fluorescence signal in the wild-type cells; a transient increase in fluorescence yield followed by a slower-phase decrease was observed. This behavior was interpreted as due to an initial electrogenic exit of ATP from the mitochondrial matrix to replenish that consumed in the hexokinase reaction, the subsequent decrease in probe fluorescence yield arising from activation of respiration. In nonrespiring cells, D-glucose induced only a decrease in fluorescence yield that was ascribed to an ATP-dependent polarization of the plasma membrane. In mutant cells lacking cytochrome *a*, cytochrome *b*, or both, the diS-C$_3$-(5) fluorescence yield was high and remained unaffected by CCCP and nystatin, suggesting that the membranes of the mutant are virtually entirely depolarized. A partial polarization, apparently associated with the plasma membrane, could be observed upon glucose addition to probe-mutant cell suspensions.

Mikeš and Dadák[61] have found that a series of cationic berberine derivatives appear to be sensitive to processes occurring on the outer surface of the inner mitochondrial membrane. Succinate oxidation in this system causes an increase in fluorescence yield that has been largely ascribed to a change in the quantum yield of the probe as it moves further into the hydrophobic region of the membrane. The largest fluorescence yield increase was observed using the 13 methyl derivative; a small increase in the amount of bound probe was also detected upon energization of the mitochondrial preparation. The fluorescence yield de-

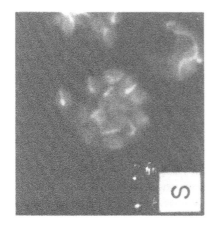

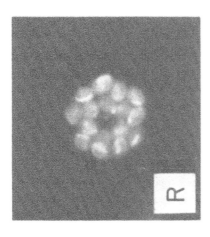

FIGURE 1. Mitochondria in *Plasmodium falciparum* stained with rhodamine 123 and visualized by fluorescence microscopy. Plates R and S illustrate segmented schizonts containing individual merozoites; the single mitochondrion in each merozoite is readily visible, especially in plate S. Plate T illustrates the loss of rhodamine 123 fluorescence when the uncoupler CCCP is present. The photographs are magnified × 3200. (From Divo, A. A., Geary, T. G., Jensen, J. B., and Ginsburg, H., *J. Protozol.*, 32, 442, 1985. With permission.)

creased linearly with the addition of subsaturating quantities of the uncoupler FCCP to the mitochondrial suspensions. The more hydrophobic berberine derivatives inhibited NAD-linked respiration, but, had no effect on succinate oxidation up to 10 μM concentration. The latter inhibitory effect has been explored in more detail by Mikeš and Yaguzhinskij.[62]

Tedeschi[8] (and references therein) has strongly questioned the use of K^+ distribution in the presence of valinomycin or other permeable ionic species to determine the magnitude of either metabolically driven membrane potentials or those of diffusion potentials used in the calibration of probe energy-dependent optical signals. This investigator has employed corrections to the matrix volume of phosphorylating mitochondria which result in a significantly smaller value of the concentration ratio of K^+ in the matrix to that in the external medium than that obtained in experiments in which the osmotic activity of this organelle has not been considered. Tedeschi[8] has further questioned the use of the Nernst equation over a wide K^+ concentration range and proposes that the observed K^+ gradient across the mitochondrial inner membrane can be explained by a Gibbs-Donnan distribution involving K^+ and impermeant anions present in this preparation.[63]

In contrast to the findings of Laris et al.[24] in hamster liver mitochondria using diS-C$_3$-(5), Tesdeschi[64] found a negligible change in the membrane potential induced by metabolism in mitochondria from *Drosophilia* using the probe CC$_6$. The authors have suggested that Mg^{++} present in the working medium employed by Laris et al.[24] cause erroneous values for the calibrating diffusion potentials to be obtained and point out that in work by Laris[42] in which ATP, but, no Mg^{++} is present, a significantly lower value for the membrane potential is obtained. It should be noted, however, that Mg^{++} is a cofactor for ATP hydrolysis and that the lower value of $\Delta\Psi$ in these experiments may reflect this requirement. In more recent work, membrane potential measurements in mitochondria using the probes diS-C$_3$-(5), merocyanine 540, ANS, and diBa-C$_4$-(5) have been undertaken by Kinnally et al.[65] under a wide variety of conditions, again using K^+ diffusion potentials calculated according to the Nernst equation with volume corrections as a calibration for the probe energy-dependent fluorescent signal. A membrane potential of at most 60 mV, and, probably significantly less was obtained in these investigations. Similar low values have also been obtained by the Tedeschi group using microelectrode techniques[8] (and references therein).

Clearly, the estimate used for the matrix volume of actively phosphorylating mitochondria is critical in probe calibration work based on diffusion potentials. The large discrepancies reported for $\Delta\Psi$ by different investigators, even when the same preparation and probe are employed, indicate the need for caution and for further clarifying studies when the Nernst equation is used in determining potential gradients from ion distributions in mitochondria suspensions.

The anionic N-phenylnapthylamines such as ANS and MNS have long been employed as fluorescent probes in mitochondrial systems. Williams et al.[66] provide a comphrehensive list of work performed with these probes in both mitochondria- and EDTA-type submitochondrial particles that lack the F$_1$ portion of the ATPase. Azzi et al.[67] have found that the introduction of substrate to mitochondria causes a two-fold increase in the Michaelis constant K_m characterizing the interaction of ANS with intact mitochondria, and in EDTA-type submitochondrial particles, a decrease of 3.6-fold in K_m is observed with no change in the total number of sites available to the probe in either system. The basis of the characteristic fluorescence changes exhibited in these preparations was explained on the basis of a change in binding constants, such that in intact mitochondria, the probe is ejected from the membrane to the bulk aqueous phase with a decrease in fluorescence yield, whereas in submitochondrial particles, the probe was transferred from the medium to the membrane with a corresponding increase in emission yield. The fluorescence yield change due to valinomycin-induced K^+ diffusion potentials was linear over an external K^+ concentration range of 100 mM. Based on this calibration and a value of 10 mM for the internal K^+ concentration, succinate oxidation, supports a membrane potential of 180 mV.

A more detailed analysis of the ANS interaction with the mitochondrial membrane by Williams et al.[66] indicates that ANS readily penetrates the inner membrane, since no additional probe binding to the membrane occurred in osmotic shock work which disrupted the organelle. Barrett-Bee and Radda[68] have further shown that in submitochondrial particles, an increase in the ANS or MNS quantum yield is observed in succinate- or NADH-energized submitochondrial particles relative to that in the inhibited or uncoupled preparation, which can also be detected as an increase in the probe fluorescence lifetime. The quantum yield change persists, however, if the probe-submitochondrial particle system is allowed to become anaerobic, but, not in inhibited or uncoupled particles. These observations have been interpreted in terms of a two-state membrane configuration model. Ferguson et al.[69] have demonstrated that although the rate at which the ANS and MNS ATP-dependent fluorescence enhancement occurs in bovine heart submitochondrial particles is independent of the total ATPase activity which was systematically decreased by Nbf-Cl (NBD-Cl) titration, this rate is sensitive to the ATPase turnover rate, since the rate of probe fluorescence enhancement is decreased when less rapidly hydrolyzed substrates such as ITP are substituted for ATP. Based on these observations, it is suggested that the rate of fluorescence yield enhancement in this preparation may not be sensitive to a delocalized energization process, but, to local phenomena occurring at the level of the ATPase. The use of these probes as indicators of membrane potential in submitochondrial particles was thus questioned.

Robertson and Rottenberg[21] have recently carefully reexamined the behavior of ANS in rat liver mitochondria and concluded that the probe has two classes of binding sites: a high-affinity, low-capacity ($k_d = 10$ to 50 μM, n = 3 to 8 nmol/mg protein) class in which bound ANS is strongly fluorescent and a low-affinity, high-capacity ($K_d > 500$ μM, n > 50 nmol/mg protein) class with little fluorescence. The observation that the apparent ANS dissociation constant depends on the inner mitochondrial membrane surface potential has been exploited to obtain estimates of the latter potential; the surface potential appears to be unchanged during the oxidation of substrates, a conclusion that does not support the role of the surface potential predicted in the models of Kell[19] and Malpress[17] for energy transduction. In agreement with Azzi et al.,[67] the reduction in fluorescence yield was found to be best explained by the expulsion of the probe from the mitochondrial membrane when $\Delta\Psi$ is formed. The potential-dependent ANS fluorescence yield changes were correlated with the $\Delta\Psi$ values, but, tended toward saturation at high values, a common observation with probes that respond to charge separation by a redistribution-type mechanism.

A number of polyene probes of the oxonol class have been extensively characterized in submitochondrial particles. Oxonol V, one of the original probes of this class, bears a negative charge at physiological pH that is delocalized over the conjugated carbon chain and part of the ring systems, since the pK of the hydroxyl proton is approximately 4. In submitochondrial particles, oxonol V exhibits a 3 to 5 nm absorption spectrum red shift when a variety of substrates such as NADH, succinate, or ATP are employed as energy sources. The red shift is accompanied by a nominally 80% loss in fluorescence yield with a minimal shift in the fluorescence emission spectrum maximum. The energy-dependent oxonol V signals are observed in a variety of energy-transducing preparations that produce an inside-positive potential gradient. In PHM, however, an oxonol V absorption spectrum blue shift is observed in the presence of ATP. Oxonol VI, the propyl derivative, exhibits a much larger absorption spectrum shift, 15 to 20 nm, but only a 20% loss in fluorescence yield when sustrate is supplied to dye-submitochondrial particle suspensions. All of these energy-dependent signals can be reversed by uncouplers of oxidative phosphorylation. In fixed-wavelength work, the absolute magnitude of both the dye absorbance and fluorescence signals is increased by either NH_4Cl or nigericin + K^+. The resulting signal can be restored to the control level by valinomycin or by permeant anions such as thiocyanate.[70,71] These observations indicate that these oxonols are specifically sensitive to the membrane potential component of the electrochemical gradient, since the increase in the absolute magnitude of the oxonol energy-dependent signals occurs under conditions in which $\Delta\Psi$ is increased at the expense of ΔpH.

In an elegant series of double-probe experiments in which oxonol VI was used to monitor $\Delta\Psi$ and 9-aminoacridine to measure ΔpH in submitochondrial particles, Bashford and Thayer[71] have determined the equilibrium phosphate potential value. This experiment involved the variation of the probes' NADH-dependent optical signal by the manipulation of the ATP/ADP xP_i ratio, i.e., the phosphate potential. It was demonstrated that the null point in this titration occurred at the same phosphate potential value for both probes. Using the equilibrium value of this potential, approximately 11 kcal/mol, the value obtained for the total electrochemical gradient was 228 to 237 mV. The contribution from the pH gradient was estimated to be 159 to 171 mV, which by subtraction gives 57 to 78 mV for the membrane potential. In low-salt medium, the membrane potential from ATP hydrolysis was estimated to be 93 to 102 mV.

The behavior of oxonol VI under both equilibrium and time-resolved conditions in bovine heart submitochondrial particles has been investigated in detail by Smith and Chance.[72] Binding analyses based on fitting optical data to the Langmuir adsorption isotherm indicate that the dye dissociation constant, K_d, decreases and the maximum number of binding sites, n, increases in the presence of ATP relative to the values of these binding constants obtained using oligomycin-inhibited particles. These observations indicate that the basis for the oxonol absorption spectrum red shifts is the transfer of probe from the aqueous medium to either membrane-binding sites or the inner volume of the vesicular preparation. In PHM, an increase in the ratio of K_d to n is observed in ATP-supplemented dye-membrane suspensions relative to that in the uncoupled preparation. The dye is thus expelled from the mitochondrial membrane under energy-transducing conditions, thereby accounting for the observed absorption spectrum blue shift in this case.

The time course of the larger oxonol VI energy-dependent absorption spectrum red shift observed in submitochondrial particles was also studied using rapid-mixing techniques. The time course of the signal is distinctly biphasic with the faster phase developing on a time scale of tenths of seconds, whereas the slower phase occurs on a scale of tens of seconds. The faster phase signal followed a pseudo-first order rate law with a second order rate constant of 3×10^5 M(dye)$^{-1}$ sec^{-1}. The slower phase rate constant was independent of oxonol VI concentration and appears to reflect a true first-order process. The faster phase of the oxonol VI signal has been interpreted as being due to the initial transfer of dye from the aqueous phase to membrane-binding sites, as suggested by the equilibrium-binding analyses. The slower phase may be due to the permeation of the membrane by this probe. The slow phase signal is observed in phospholipid vesicles only under conditions in which the bilayer is permeable to this dye.[73] The rate at which the ATP-dependent cytochrome c oxidase soret band shift develops and the amplitude of this signal are decreased by the presence of oxonol VI[72] suggesting that this probe does permeate the submitochondrial particle membrane by an elecrophoretic process that competes with that responsible for the soret band shift signal for energy from the electrochemical gradient. The second-order rate constant characterizing the faster phase of the oxonol VI energy-dependent signal was the same when ATP, NADH, or oxygen pulses to an anaerobic dye-submitochondrial particle system were used to produce the dye absorption spectrum red shift. This observation suggests that by this kinetic criterion, the oxonol VI response is not dependent on a local enzymatic activity and in this sense is due to a delocalized charge gradient. (See the work of Kiehl and Hanstein[74] on complex V using this dye, however, that is described later in this section.)

Smith et al.[75] have further demonstrated that oxonols V and VI undergo uncoupler-sensitive, ATP-dependent fluorescence depolarization and a decrease in the lifetime of the longer-lived probe species in bovine heart submitochondrial particles. These findings are consistent with the formation of regions of high local dye concentration, either within the membrane or by accumulation of this probe in the internal volume of this vesicular preparation. Oxonol V also exhibits Stern-Volmer behavior in solution, i.e., the reciprocal of the probe lifetime is a linear function of the probe concentration in ethanol; the probe is thus self-quenched in this system.

Oxonol V appears to be sensitive to the onset of ischemia in the perfused rat heart. In experiments employing flying spot surface fluorimetry,[76] the control dye fluorescence histogram becomes bimodal under conditions of partial ischemia; the latter histogram then becomes unimodal and shifts to lower intensity when complete ischemia is imposed on the heart.[6] These observations suggest that oxonol V may be a useful indicator of membrane potential in this preparation, although the mitochondrial contribution to the loss in fluorescence yield observed in this preparation has yet to be established.

Sanadi et al.[77] and Pringle and Sanadi[78] have recently employed oxonol VI to monitor membrane potential formation in the mitochondrial ATPase reconstituted in vesicles. The investigators have presented evidence that coupling factor B (F_B) is an essential component of the proton channel of F_O, since modification of F_B by Cd^{++} inhibits proton conduction through the channel.

Kiehl and Hanstein[74] have compared the behavior of oxonol VI in submitochondrial particles with that in a mitochondrial ATPase (complex V) preparation that was not reconstituted in phospholipid vesicles. Among the differences in the probe response noted by these investigators in these two preparations is the persistence of an uncoupler-sensitive oxonol VI absorption spectrum red shift elicited by ATP in complex V preparations containing high quantities of cholate or desoxycholate detergent that are known to be sufficient to disrupt the membrane integrity in submitochondrial particles. Since the ATP-induced probe optical signal observed in the complex V preparation can also be fully restored to the control level by valinomycin, the possibility that either vesicle structures in this preparation are more resistant to detergents, presumabaly because they would contain less phospholipid or that a volume with solvent characteristics formed solely by the proteins constituting the enzyme system similar to that suggested by Wagner et al.[79] for CF_1 is present in complex V. Such a molecular vesicle would have to be impermeable to thiocyanate, which is essentially without effect on the oxonol VI red shift in this preparation, and resistant to high concentrations of detergents, but, be permeable to valinomycin and the probe itself. In the absence of hard evidence for the existence of such an unusual structure, the possibility that the energy-dependent oxonol VI optical signal is due to local charge effects, instead of a presumed transmembrane potential gradient, must also be considered. A reduction in the negative surface charge density by the association of protons or other cations would be expected to promote additional association of the anionic probe with the membrane which would produce a red shift in the probe absorption spectrum. In sarcoplasmic reticulum, for example, the presence of Ca^{++} promotes the binding of this probe to the preparation membrane. Since the investigations with oxonol VI in the complex V preparation were carried out in the presence of 10 mM $MgSO_4$, however, the membrane surface charge in this preparation is probably sufficiently screened to render surface potential changes a rather minor contributor to the oxonol VI optical signal. That the oxonol probe is in part bound to the F_1 portion of the ATPase and that the probe-binding site properties are altered by ATP when it binds to the complex may also be significant in this study. Significantly, the investigators observed that the magnitude of the oxonol VI red-shift signal was a function of the ATP/ADP ratio, rather than the phosphorylation potential, as observed by Bashford and Thayer[71] in submitochondrial particles. Thus, the probe signal may in part result from processes originating at the level of the ATPase in complex V. The kinetic results obtained with a number of different substrates by Smith and Chance[72] that have been previously described, however, argue against a significant local contribution to the probe signal in phosphorylating submitochondrial particles.

The interaction of merocyanine 540, which bears a localized negative charge with bovine heart submitochondrial particles has been investigated by Smith et al.[80] Detailed binding studies, based on the methods of Bashford and Smith[29] that have been previously described, indicate that a substantial decrease in the dye dissociation constant occurs in the presence of ATP relative to that characterizing the oligomycin-inhibited preparation. Little or no change in the maximum number of binding sites was observed in this system when energy-transduction was initiated by ATP addition; these findings suggest that no significant alter-ation in membrane structure at the level of the sites available to M540 has occurred as a consequence of charge separation in the submitochondrial partical preparation. These results suggest that the origin of the dye energy-dependent signal in this membrane system is a redistribution of probe from the aqueous phase to membrane-binding sites. The absolute magnitude of this signal is increased by either nigericin + K$^+$ or NH$_4$Cl, indicating that M540 is specifically sensitive to the membrane potential portion of the electrochemical gradient.

In contrast to the millisecond response times of the M 540 fluorescence changes in excitable preparations such as the giant axon from the squid *Loligo peali* (Cohen et al.[81]) or in frog heart preparations (Salama and Morad[82]) during the action potential development, the char-acteristic probe absorption spectrum changes developed on a time scale of tens of seconds, significantly slower than those of oxonol VI in the submitochondrial particle preparation. The time course of the energy-dependent signal was monophasic and could be fit to a single exponential rate equation. The rate constants were a function of dye concentration, however; a second-order rate constant value of nominally 1×10^4 M(dye)$^{-1}$ sec^{-1} was obtained with either succinate, NADH, or ATP as the energy source. The results from the kinetic studies suggest that M540 can respond to charge separation in the mitochondrial preparation only as fast as it can redistribute from the aqueous medium to the membrane; no evidence for a faster process was detected in the rapid-mixing techniques used in these measurements.

Unlike the effect of the oxonols and cyanines previously discussed on energy-requiring processes in either submitochondrial particles or intact mitochondria, M540 has no effect on the rate at which the ATP-dependent cytochrome *c* oxidase soret band shift signal develops, nor on the rate at which the oxidation of cytochrome *c* or the formation of NADH, both by ATP-dependent reversed electron transport, occurs. These observations suggest that M540 does not permeate the submitochondrial particle membrane on the 0.3 sec to 20 min time scale covered by these three observations. The M540 binding constants and the rate at which the energy-dependent optical signal develops are both quite sensitive to the medium ionic strength. The probe appears to be confined largely to the surface of the membrane and may be at least in part responding to changes in surface potential, possibly occurring by the mechanisms suggested by the theoretical work of Malpress[17] and Kell[19]. In experiments with M540 in black lipid membranes, Krasne[83] has also found that no measurable current can be ascribed to the translocation of this probe when a potential gradient across the bilayer is present. Russell et al.[84] have also suggested that M540 is sensitive to electrostatic surface potential changes in sarcoplasmic reticulum vesicles caused by the association of cations with the vesicle membrane, but, not to the presence of diffusion potentials (See the fol-lowing.). No work on calibrating the M540 probe response has yet been attempted in the submitochondrial particle system, but, in view of the preceding findings, the results of such experiments will be of considerable significance to the study of surface potential effects in this system, since the probe does not appear to distribute across the membrane in accordance with prevailing potential gradients.

Table 1
MITOCHONDRIA AND RELATED PREPARATIONS

Author(s)	Probe(s)	Ref.
Adamski, Morkowski and Michejda	diS-C$_3$-(5)	40
Akerman	safranine	45
Akerman and Wikström	safranine	26
Azzi, Gherardine, and Santato	ANS	67
Bakeeva, Kirillova, Kolesnikova, Konoshenko, and Mokhova	diS-C$_3$-(5)	38
Bammel, Brand, Germon and Smith	diS-C$_3$-(5)	28
Barret-Bee and Radda	ANS, MNS	68
Bashford, Barlow, Chance, Smith, Silberstein, and Rehncrona	oxonol V	6
Bashford, Chance, Smith, and Yoshida	oxonol V	73
Bashford and Smith	oxonol V	29
Bashford and Thayer	oxonol VI	71
Bereiter-Hahn	DASPMI	52
Chance	oxonol V (MC V)	142
Chance, Baltscheffsky, Vanderkooi, and Cheng	merocyanine 540 (MC I), merocyanine II (MC II), AS, oxonol V (MC V)	5
Cockrell	safranine	50
Conover and Schneider	diO-C$_2$-(5), diO-C$_6$-(5), diS-C$_2$-(5), pinacyanol, ethyl red, merocyanine 540, safranine O, and others	32
Davis, Weiss, Wong, Lampidis, and Chen	rhodamine 123	58
Divo, Geary, Jensen, and Ginsburg	rhodamine 123	59
Ferguson, Lloyd, and Radda	ANS, MNS	69
Johnson, Walsh, Bockus, and Chen	rhodamines 3B, 6G, 123, diO-C$_2$-(3), diO-C$_4$-(3), diO-C$_5$-(3), diO-C$_6$-(3), diO-C$_2$-(5), safranine O	57
Johnson, Walsh, and Chen	rhodamine 123	56
Kauppinen and Hassinen	safranine	48
Kiehl and Hanstein	oxonol VI	74
Kim, Kudzina, Zinchenko, and Evtodienko	diS-C$_3$-(5)	51
Korchak, Rich, Wilkenfeld, Rutherford, and Weissmann	diO-C$_6$-(3)	55
Kovac and Varecka	diS-C$_3$-(5)	60
Laris	diS-C$_3$-(5)	42
Laris, Bahr, Chaffee	diS-C$_3$-(5)	24
Lauquin, Villiers, Michejda, Brandolin, Boulay, Cesarini, and Vignais	diS-C$_3$-(5)	43
Mewes and Rafael	DSMP	53
Michejda, Adamski, Hejnowicz, and Hryniewiecka	diS-C$_3$-(5)	39
Mikeš and Dadák	berberines	61
Montecucco, Pozzan, and Rink	diS-C$_2$-(5), diS-C$_3$(5), diO-C$_2$-(3), diI-C$_6$-(5)	34
Pechatnikov, Rizvanov, and Turchina	diS-C$_2$-(5), diS-C$_3$-(5)	33
Pringle and Sanadi	oxonol VI	78
Quintanilha and Packer	diS-C$_3$-(5)	36
Rafael and Nicholls	DSMP	54
Riley and Pfeiffer	safranine	49
Robertson and Rottenberg	ANS	21
Sanadi, Pringle, Kantham, Hughes, and Srivastava	oxonol VI	77
Smith and Chance	oxonol VI	72
Smith, Frank, Bashford, Chance, and Rudkin	diBA-C$_4$-(5), diS-C$_3$-(5), diS-C$_5$-(5), oxonol V, oxonol VI, merocyanine 540	146
Smith, Graves, and Williamson	merocyanine 540	80
Smith, Halliday, and Topp	oxonols V, VI, VII, VIII	75
Smith, Powers, Prince, Chance, and Bashford	oxonols V, VI, VII, VIII, IX	7
Smith, Russ, Cooperman, and Chance	oxonol V	70
Tedeschi	CC$_6$	64

Table 1 (continued)
MITOCHONDRIA AND RELATED PREPARATIONS

Author(s)	Probe(s)	Ref.
Walsh Kinnally, Tedeschi, and Maloff	ANS, diBA-C_4-(5), diS-C_3-(5), merocyanine 540	65
Wikström and Saari	cytochrome aa_3	10
Williams, Layton, and Johnson	ANS	66
Wilson	ANS and safranine	47
Zanotti and Azzone	safranine	46
Zinchenko, Holmuchamedov, and Evtodienko	diS-C_2-(5),diS-C_3-(5)	27

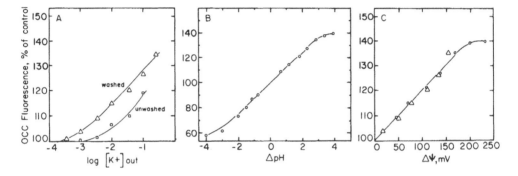

FIGURE 2.   Titration curves relating the fluorescence signal of the probe to (A) the K^+ concentration, (B) to proton, and (C) to potassium diffusion potentials. ΔpH and $\Delta\Psi$ were calculated from the Nernst equation. Note that the plot in (B) tends toward saturation near both extrema, whereas, a saturation effect is present in (C) above approximately 175 mV. (From Pick, U. and Avron, M., *Biochim. Biophys. Acta*, 440, 189, 1976. With permission.)

IV. PHOTOSYNTHETIC SYSTEMS

In early work with the oxacarbocyanine probe OCC in chromatophores from *Rhodospirillum rubrum,* Pick and Avron[85] demonstrated a reversible light-induced probe fluorescence yield increase that was sensitive to electron transport inhibitors, permeable ions, and uncouplers of oxidative phosphorylation. The reduction in the light-dependent OCC fluorescence yield increase caused by permeable ions was accompanied by a quenching of 9-aminoacridine on a similar time scale, suggesting that a compensating increase in ΔpH occurs as $\Delta\Psi$ is reduced. In the chromatophore preparation, OCC fluorescence-yield changes in the dark could be induced by the addition of permeable ions or by rapid pH alterations. Assuming that these changes represent diffusion potentials, calibration curves that were linear from 0 to 200 mV and over ± 4 ΔpH units from zero could be obtained; typical plots are illustrated in Figure 2. A maximum value of the light-induced $\Delta\Psi$ of 170 mV and a steady-state value of 90 to 110 mV were obtained. These values contain a correction of nominally 20 mV, since OCC itself was found to reduce the value of the probe light-induced signal. The probe response to diffusion potentials generated in the dark was found to be sensitive to the oxidation state of the electron carriers and to the presence of strong oxidants and reductants, indicating that the calibration curves must be used with caution in estimating the light-induced $\Delta\Psi$, since changes in the redox state of these carriers occurs upon illumination of the preparation. The reported $\Delta\Psi$ values, however, agree with those obtained using the SCN^- distribution technique, but, are considerably smaller than those estimated from carotenoid band shift signals. The latter signals, however, have been found to contain a significant surface potential component in related preparations; see what follows. The difference absorption spectrum of OCC in the chromatophore system was nearly identically reproduced by that of OCC in a *sec*-butyl acetate/*n*-hexane (3:1, v:v) mixture relative to OCC in water.

Schuurmans et al.[86] have employed oxonol VI in spinach thylakoid preparations using dual-wavelength spectroscopy. Under continuous actinic illumination, the proton uptake signal was accompanied by a rapid decrease in the dye absorbance at 590 nm, which then incresed to a steady-state level that persisted as long as illumination was in effect. The illumination-induced difference spectrum of the probe indicated that a spectral shift had occurred as in submitochondrial particles and that this shift could also be demonstrated by the formation of valinomycin-K^+ diffusion potentials. Using the latter potentials as calibrations, the initial response of the oxonol VI dye was estimated to correspond to a membrane potential of 90 mV that decayed to 50 mV under steady-state illumination. In single-turnover flash work, the dye response developed within 20 msec and when compared to the carotenoid bandshift signal, the onset of the dye response approximately coincided with the faster decay phase of the carotenoid signal, whereas the decay of the probe signal paralleled the slower phase decay of the carotenoid response. It was concluded that oxonol VI is capable of detecting the initial production of $\Delta\Psi$, which was estimated to be 50 mV in the flash excitation experiments.

The preceding studies have been extended by Admon et al.[87] to demonstrate that nigericin enhances the magnitude of both the transient and steady-state oxonol VI absorbance signal observed upon illumination of lettuce leaf chloroplasts. The light-induced probe signal was also better maintained when P_i was present in the reaction medium and was increased when the temperature was lowered, presumably due to reduced membrane proton permeability at the lower temperature. In the dark, an ATP-induced oxonol VI signal similar to that observed upon illumination of the chloroplast preparation could be observed, provided that it had been activated by preillumination. The light- and ATP-dependent oxonol VI signals were reversed by nigericin plus valinomycin, the uncoupler SF6847, and by a number of ionophores and inhibitors, including gramacidin, tentoxin, Dio-9, and DCCD. These observations, which are in agreement with the results of Shuurmans et al.[86] indicate that a substantial, but transient, potential gradient is formed upon illumination of the chloroplasts or by the addition of ATP in the dark to light-activated preparations. The presence of relatively large $\Delta\Psi$ transient offers an explanation for early-phase photophosphorylation and luminescence from the photosystem II primary acceptor, Q, via ATP-driven reversed electron transport. K^+ diffusion potentials could not be used to calibrate the oxonol VI energy-dependent signal, because this probe is sensitive to ionic strength changes caused by KCl addition (See the following pages.). An attempt to calibrate the signal was made using acid addition to the external medium; with the resulting dye signals as approximate calibrations, a steady-state $\Delta\Psi$ of 50 mV was estimated. The acid-induced probe signals, however, were not sensitive to uncouplers, suggesting that the $\Delta\Psi$ value derived from these experiments must be treated with caution.

In related work, Shahak et al.[88] have reconstituted the CF_0-CF_1 ATPase from spinach chloroplasts in soybean phospholipids by three procedures. ATP addition caused a change in the 603 to 630 nm absorbance signal of oxonol VI that could be incremented by nigericin addition and completely reversed by subsequent addition of valinomycin. The ATP-induced signal increased as the protein to lipid ratio was incremented; the corresponding $\Delta\Psi$ was estimated to be 40 to 50 mV using calibration methods to be described subsequently. Lowering the temperature decreased the rate at which the probe signal developed, but led to a slight increase in the final ATP-induced absorbance change, as expected from a decrease in both the ATPase turnover number and membrane proton permeability. Due to the ionic strength sensitivity of oxonol VI, K^+ diffusion potentials could not be used to calibrate the probe response. A pH jump procedure using H_2SO_4, however, could be employed to generate proton diffusion potentials that completely decayed to the control level over time and could be reversed by valinomycin addition. The ATPase vesicles prepared by a cholate dilution procedure were too leaky for proton diffusion potentials to be detected, but a dye response to acid injection could be detected when some of the cholate was removed by Sephadex® chromatography. The vesicles formed by a freeze-thaw-sonicate procedure appeared to have the lowest permeability, since the acid jump-induced probe signal decayed on a time scale of minutes. The oxonol VI absorbance signal due to acid injection increased as the temperature was lowered, and at fixed temperature, the probe signal decay rate increased as the protein to lipid ratio was incremented.

Shahak and Pick[89] have demonstrated a lag between the onset of ^{32}P incorporation into ATP and ATP hydrolysis in the CF_0-CF_1 complex reconstituted in soybean phospholipids. At 37°C, the lag persists for some 2 min, but, can be reduced to zero by raising the temperature to 40°C. A similar lag of 20 sec duration has also been observed in light-triggered ATP-P_i exchange in intact chloroplasts. The ATP-P_i exchange kinetics, however, do not appear to correlate with the formation of $\Delta\Psi$, as monitored by oxonol VI. A conformation change in the ATPase complex, initiated by ATP hydrolysis, that is a prerequisite for the exchange process to be catalyzed has been suggested as the source of the lag between ATP hydrolysis and the exchange phenomenon.

De Wolf et al.[90] have employed neutral red and cresol red in studies of well-coupled photosystem I-enriched subchloroplast vesicles under single-turnover conditions. A flash-induced bleaching of the 550 nm neutral red absorbance signal was observed to occur in less than 5 msec, that was followed by a slow phase absorbance increase at this wavelength. The fast-phase bleaching was not sensitive to buffers, ionophores, or uncouplers. The origin of this signal was suggested to be the reduction of neutral red on the vesicle membrane surface due to an increase in membrane negative surface charge that would attract the protonated neutral red species through coulombic interactions. This interpretation, however, is at variance with that of Junge et al.,[91] see what follows. The corrected slow phase neutral red signal was readily eliminated by buffers, valinomycin, or uncouplers and was ascribed to the protonation of the probe when it functions as a pH indicator. No flash-induced proton translocation could be detected by neutral red or cresol red. These observations were interpreted as compatible with a model proposed by Schurrmans et al.,[92] in which vectorial proton translocation is not a prerequisite for energy transduction under single-turnover conditions. The preceding interpretation of the transient decrease in the neutral red absorbance signal is in marked disagreement with that of Thieg and Junge[93] who have concluded from a series of investigations on chloroplast preparations that this probe functions as a pH indicator and that the uncoupler-sensitive loss of the neutral red signal reflects the buffering of protons, derived from water oxidation, inside thylakoids by compartmentalized regions with limited buffering capacity, since the reduction in the neutral red absorbance could be reversed by repeated actinic light flashes. Further support for this interpretation was derived from the observation that the time course of the reduction in the neutral red signal did not correlate with that of the transmembrane potential development monitored, using the absorbance of an electrochromic complex at 522 nm, thereby suggesting that the protons had not crossed the thylakoid membrane.

In related work with spinach and pea chloroplasts in which the CF_0 portion of the ATPase was extracted by EDTA treatment, Junge et al.[94] have monitored the electrical potential difference and proton production from water oxidation using electrochromic signals from a lutein-chlorophyll *b* complex and neutral red, respectively, via time-resolved absorption spectroscopy. The extraction procedure causes a transient loss of protons from water oxidation and a 10% reduction in the electrical potential difference. These observations were interpreted as indicating direct access to the CF_0 channel by these protons, plus an enhanced CF_0 hexacooperative buffering capacity, due to proton-binding groups in the channel, induced by EDTA treatment, that accounts for the "loss" and the eventual reappearance in the external phase of these protons via passage through the CF_0 channel.

Peters et al.[95] have extensively examined the behavior of oxonol VI and the carotenoids in Photosystem I-enriched subchloroplast vesicles. Under conditions in which redox potentials were carefully poised for optimum electron transport,[96] the oxonol and the slow phase carotenoid signals exhibited similar, but not idential, time courses; these signals could be abolished by saturating preillumination of the vesicles or by a preestablished transmembrane pH gradient, the fast component of the carotenoid signal only being observed under these conditions. The slower phase carotenoid and oxonol VI signals were significantly more sensitive to the presence of valinomycin, various uncouplers, and to digitonin than was the faster phase signal. Under optimum cyclic electron transport conditions, nigericin was without effect on these signals, but at pH 8, a suboptimum pH, nigericin did enhance the signal, indicating that under the latter conditions proton displacement is likely implicated in the probe response to illumination of this preparation, possibly at membrane sites near the bulk phase-membrane interface. The effect of the various perturbants also suggests that the origin of the slow phase carotenoid and oxonol VI signals is at or near the membrane surface. Differences in the activation energies characterizing the carotenoid and oxonol VI response mechanisms imply a somewhat different location in the membrane for these two probes. During continuous pulsed illumination of the vesicles, no changes in the oxonol VI fluorescence lifetime or a change in probe membrane-binding was detected, suggesting that the oxonol response is not self-limited by the electrophoretic translocation of the probe across the bulk membrane, as suggested by Galmiche and Girault[97] based on the observation that either valinomycin + K^+ or oxonol VI inhibited ATP synthesis in flash illumination experiments, but, not under conditions of continuous illumination. The latter investigators demonstrated that the oxonol VI response to either illumination or ATP hydrolysis in intact spinach chloroplasts was qualitatively the same and was not dependent on the magnitude of ΔpH. It should be noted that in the work by Galmiche and Gerault the ratio of dye to membrane (nmol dye/μg chl) was 250 to 500 times that used by Peters et al.; the likelihood of the availability of free oxonol VI to redistribute onto and across the chloroplast preparation membrane is thus much larger in the experiments of the former than in that of the latter investigators.

Bashford et al.[98] have also employed oxonol VI in chromatophores from the photosynthetic bacteria *Rhodospirillum rubrum* S-1 and *Chromatium vinosum*. The light-induced dye absorbance decrease at 587 nm, an isosbestic point in the carotenoid band shift signal, was found to be enhanced in absolute magnitude by nigericin and abolished by valinomycin + K^+. The dye thus appears to be specifically sensitive to $\Delta \Psi$ in these preparations. The presence of oxonol VI was found to accelerate the decay of the carotenoid band shift signal; the dye response, however, was always slower than that of the carotenoids and followed a second-order rate law with a rate constant of $2 \times 10^6 M(\text{dye})^{-1} \sec^{-1}$. The oxonol VI signal was linear, with membrane potentials calculated from the intrinsic carotenid signal. In contrast to the suggestion of Peters et al., the investigators have proposed that oxonol VI does redistribute across the membrane, presumably in accordance with $\Delta \Psi$. The concentration of the probe employed in this work, however, was some five times higher than that used by Peters et al.[95]

The phosphate potential has been determined by Bashford et al.[6] using the null point titration procedure previously described[71] in chromatophores from *Rhodopseudomonas spheroides* with oxonol VI as the membrane potential indicator. At 80% of actinic light saturation intensity, a value of 11 Kcal/mol is reported, whereas, at saturation the chromatophores support a potential of about 14 Kcal/mol.

Koyama et al.[99] have used Raman spectroscopy in a study of the behavior of neurosporene, the dominant carotenoid present in chromatophores prepared from the G1C mutant of *Rho-*

dopseudomonas sphaeroides. For excitation at 472.7 nm, the ratio of the two most prominent Raman band intensities was very sensitive to small shifts in the carotenoid absorption spectrum shift caused by diffusion potentials, or by treatment of the chromatophores with oligomycin, FCCP, or deoxycholate. The intensity ratio was also temperature-dependent and appeared to contain a signal component, due to the excitation beam acting as an actinic light source.

Johnson et al.[100] have incorporated the carotenoid β-carotene into a phospholipid vesicle preparation containing bacteriorhodopsin from *Halobacterium halobium.* A light-induced membrane potential that developed within 14 to 22 μsec was detected, using the technique of kinetic Raman spectroscopy. It was shown that the membrane potential-sensitivity depended on the excited-state properties of the carotenoid, i.e., that this probe was responding electrochromically to charge separation in this preparation. This work on one of the few examples in which the electrochromic characteristics of the carotenoids have been preserved in reconstituted systems.

Szalontai[101] has employed resonance Raman measurements to monitor the formation of the membrane potential in R_1M_1 mutants of *Halobacterium halobium.* The light used in Raman measurements also served to drive the photochemical cycle of bacteriorhodopsin. The light-induced red shift in the absorption bands of the carotenoids present in the membranes of these cells causes intensity changes in the peaks constituting the Raman spectra associated with the 482.5 and 586.1 nm absorption regions. By taking the ratio of the intensity of the Raman bands indentified by wave numbers in Figure 3 to that in a spectral region where no signal is present, 1400 cm^{-1} or 1050 cm^{-1}, scattering artifacts from the whole-cell suspensions could be minimized. Analysis of the signal time course associated with the 482.5 nm absorption band yielded a time constant of 80 (± 20) μsec for potential formation; this value is considered to be an upper limit for the $BR_{570} \rightarrow M_{412}$ transition and is similar to the value of 50 μsec reported for this transition in purple membrane fragments.[102]

Ehrenberg et al.[103] have employed the styryl potential-sensitive probe RH-160 in an investigation of light-activated bacteriorhodopsin from *Halobacterium halobium* incorporated in liposomes. The RH-160 fluorescence signal, corrected for an inner filter effect, developed within nominally 20 μsec after flash illumination of the sample and in H_2O was characterized by a halftime of 38 (± 5) μsec, Figure 3. The M_{412} intermediate developed with a significantly longer lifetime of 68 (± 5) μsec and exhibited a large isotope effect. In D_2O, the M_{412} halftime was 300 (± 5) μsec, and the signal developed after a 60 μsec delay, following sample flash excitation. Charge separation thus appears to precede the formation of the M_{412} intermediate and may accompany the early K_{590} to L_{550} transition or the later conversion of L_{550} to X_{480}, a new intermediate, that then generates the M_{412} species.

Packam et al.[104] have investigated the issue of whether the light-induced carotenoid band shift signal in chromatophores from *Rhodopseudomonas capsulata* is due to local vs. delocalized membrane potentials. Under conditions in which cytochrome c_2 is available for oxidation by the reaction center, the decay of the carotenoid band shift signal generated by short light flashes is very slow, halftime of approximately 10 sec. Decay kinetics are much faster, halftime of 1 sec, when cytochrome c_2 is unavailable for oxidation, and the decay can be accelerated 20-fold by *ortho*-phenanthroline; the latter behavior was interpreted as due to a back reaction involving the reduction of the oxidized chlorophyll dimer. In each of the preceding cases, the carotenoid signal, but, not the electron transport process could be accelerated by submicromolar concentrations of FCCP or valinomycin, suggesting that the initial charge separation in this preparation is rapidly delocalized and can then be dissipated by the previously described back reaction or by electrophoretic ion flux. Under conditions in which cytochrome c_2 is unavailable for oxidation, the FCCP-sensitive proton efflux process could be decreased by valinomycin + K^+, indicating that dimer $\rightarrow$ Q electron transfer generates a delocalized membrane potential, as sensed by the carotenoids, even though electron transport may proceed across only part of the membrane dielectric.

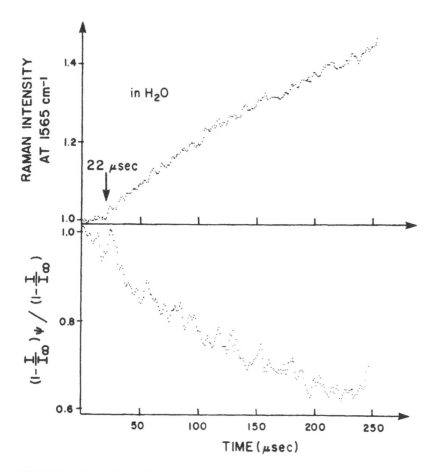

FIGURE 3.   (Upper Panel) The variation in the Raman scattering intensity at 1565 cm^{-1} of bacteriorhodopsin incorporated in lipid vesicles as a function of time over a 250 μsec 0.15 W flash at 457.9 nm. The spectral width of the detector was 10 cm^{-1}. (Lower Panel) The net effect of the membrane potential on the RH-160 panel probe fluorescence intensity, obtained by correcting the dye emission for an inner-filter effect, plotted as a function of time. A comparison of the data in these two panels suggests that the formation of a membrane potential precedes the production of the M_{412} intermediate. See the text and the indicated reference for details. (Reproduced from *Photochem. Photobiol.*, 39, Ehrenberg, B., Meiri, Z., and Loew, L. M., A microsecond kinetic study of the photogenerated membrane potential of bacteriorhodopsin with a fast responding dye, copyright 1984, Pergamon Press Ltd.)

In related work, Webster et al.[105] have shown that the carotenoid associated with the B-800-850 light-harvesting complex in *Rhodopseudomonas capsulata* is the one that responds electrochromically to transmembrane electric fields.

Matsuura et al.[22,106] have demonstrated that a carotenoid band shift signal in cells, spheroplasts, and chromatophores of *Rhodopseudomonas sphaeroides* can be observed upon addition of various monovalent or divalent salts and/or by changes in the medium pH under conditions where the transmembrane potential is controlled by an uncoupler. The salt- and pH-dependent signals were ascribed to a change in the surface potential which was estimated to be as high as 90 mV, under the conditions of these investigations. Symons et al.[107] have also estimated the contribution of surface potential changes to the carotenoid signal developed by the application of a K$^+$ diffusion potential across the membrane of chromatophores from *Rhodopseudomonas capsulata* to be 20%; the surface potential contribution could be corrected for by using Na$^+$ to maintain a constant ionic strength when the calibrating diffusion

Table 2
PHOTOSYNTHETIC SYSTEMS

Author(s)	Probe(s)	Ref.
Admon, Shahak, and Avron	oxonol VI	87
Bashford, Chance, and Prince	oxonol VI	98
Cirillo and Gromet-Elhanan	$_2$ANS	108
DeWolf, Groen, van Houte, Peters, Krab, and Kraayenhof	neutral red	90
Dutton, Prince, Tiede, Petty, Kaufmann, Netzel, and Rentzepis	bacteriochlorophyll	147
Ehrenberg, Meiri, and Loew	RH-160	103
Galmiche and Girault	oxonol VI	97
Johnson, Lewis, and Gogel	β-carotene	100
Junge, Hong, Qian, and Viale	lutein-chlorophyll *b* complex	94
Koyama, Long, Martin, and Carey	neurosporene	99
Masamoto, Matsura, Itoh, and Nishimura	NK22722, NK2273, and NK2274	23
Matsuura, Masamoto, Itoh, and Nishimura	carotenoids	22, 106
Packam, Greenwood, and Jackson	carotenoids	104
F. Peters, Van Der Pal, R. Peters, Vredenberg, and Kraayenhof	oxonol VI	95
Peters, van Spanning, and Kraayenhof	oxonol VI	96
Pick and Avron	OCC	85
Schuurmans, Casey, and Kraayenhof	oxonol VI	86
Shahak, Admon, and Avron	oxonol VI	88
Shahak and Pick	oxonol VI	89
Symons, Nuyten, and Sybesma	carotenoids	107
Szalontai	carotenoids	101
Thieg and Junge	chlorophyll *b*	93
Webster, Cogdell, and Lindsay	B-800-850, carotenoids	105

potential was developed. Additional corrections were shown to be necessary for pH effects, possibly due to H^+ diffusion potentials, as well as for suboptimal membrane to valinomycin ratios; otherwise an error approaching a factor of two in the estimation of the transmembrane potential gradient could result.

Masamoto et al.[23] have compared the illumination-induced response of the essentially impermeable merocyanines NK2272, NK2273, and NK2274 with the carotenoid band shift in *Rhodopseudomonas sphaeroides* chromatophores; structural formulae are provided in the indicated reference. The dye signal was found to be linearly related to the carotenoid signal, the salt-induced merocyanine optical signals were small in comparison to those generated by illumination, suggesting a minimal change in the chromatophore surface potential, a conclusion at variance with the theory of Malpress.[17] Significantly, the response of both the merocyanines and the carotenoid signal to uncouplers was similar. Since the uncouplers are expected to accelerate the equilibration of protons between the bulk aqueous phase and the unstirred layer near the chromatophore membrane surface, the latter observation is not in accord with the model suggested by Kell,[19] which is dependent on the existence of proton gradient between the bulk phase and the region near the membrane.

Cirillo and Gromet-Elhanan[108] have used both the distribution of NH_4^+ and SCN^-, as well as the molecular probes 9-aminoacridine and ANS to monitor ΔpH and ΔΨ, respectively, in chromatophores from *Rhodospirillum rubrum* under saturating actinic light conditions. The steady-state values of the electrochemical gradient obtained using these two methods were compared with the phosphorylation potential measured under the same conditions as ΔpH and ΔΨ to obtain the H^+/ATP ratio which was always greater than 2 and fell in the range of 2.4 to 3.4, depending on the methods used to estimate the electrochemical gradient. The difference between the H^+/ATP ratio obtained by these investigators and the higher values obtained by Kell et al.[109] has been explained on the basis of the use of less than saturating light levels by Kell et al., since the value of ΔpH supported by these chromatophores is known to be light-intensity dependent.

V. SARCOPLASMIC RETICULUM

Zimniak and Racker[110] have incorporated the Ca-ATPase into egg yolk phosphatidylcholine vesicles, which are considerably less permeable than the sarcoplasmic reticulum membrane to small ions, in order to address the issue of whether the ATP-dependent Ca^{++} uptake catalyzed by this enzyme is electrogenic. To facilitate Ca^{++} uptake, the vesicles contained a Ca^{++} trapping agent, either phosphate or oxalate. Using high concentrations of K^+, membrane potentials could be developed by valinomycin addition. These potentials were monitored by ANS fluorescence and calibrated from valinomycin-K^+ diffusion potential data from vesicles containing no ATPase. The stimulation of Ca^{++} uptake by valinomycin in the reconstituted system was studied as a function of the K^+ membrane potential under voltage clamp conditions in which the Ca^{++} translocation did not affect $\Delta\Psi$. The data were extrapolated to the potential at which no valinomycin stimulation of Ca^{++} uptake was observed. This potential, $+51$ mV, was considered to be characteristic of ATP-stimulated Ca^{++} uptake in the vesicle preparation, but, not necessarily of the enzyme itself, i.e., Ca^{++} translocation is inherently electrogenic in this system. In an independent set of experiments, ATPase-containing vesicles were prepared with K^+ in the vesicle internal volume at varying concentrations. ATP-dependent Ca^{++} uptake was measured as a function of the membrane potential developed by valinomycin addition. The $\Delta\Psi$ value at which no valinomycin stimulation was observed in this case was $+61$ mV in good agreement with that derived from the work with the ANS probe. In the sarcoplasmic reticulum membrane, however, any membrane potential developed by the ATPase activity may be rapidly dissipated by compensating small ion translocation, although a transient $\Delta\Psi$ could result from Ca^{++} translocation; see the following pages.

Young et al.[111] have employed diO-C_5-(3), a positively charged cyanine in work with type II reconstituted sarcoplasmic reticulum vesicles, which were demonstrated to lack a functional Na^+, K^+ channel. Without valinomycin, the dilution of K^+-loaded vesicles into a Na^+ medium produced no optical probe signal, but, when the ionophore was present, an inside-negative potential could be detected by alterations in the probe fluorescence yield, which persisted for some 15 or more minutes, the decay rate apparently reflecting the slow influx of Na^+ or the TRIS cation. Meissner and Young[112] have used diO-C_5-(3) to monitor proton diffusion potentials in these vesicles and found that both types of vesicles readily develop proton gradients without the presence of protonophores. It was concluded that a H^+ transport system probably exists in both types of vesicles that is independent of the Na^+, K^+ channel. In similar work, Pechatnikov et al.[113] have employed diS-C_3-(5) in demonstrating that Type I vesicles are permeable to Na^+ and K^+, where Type II vesicles can support diffusion potentials involving these ions that are detectable by the probe. Conductivity coefficient estimates are provided and a physiological function suggested for the Na^+ and K^+ permeability pathways. The Pechatnikov group[114] has also reported studies of the sarcoplasmic reticulum membrane permeability based on chloride diffusion potentials simultaneously monitored by diS-C_3-(5) and a laser light-scattering technique. At concentrations higher than those usually employed in these preparations for potential gradient measurements, Pechatnikov et al.[115] however, have found that diS-C_3-(5) increases the permeability of the sarcoplasmic membrane in a manner similar to that caused by detergents of temperature increases. The probe diO-C_2-(5) has been employed by Ohnishi[116] in measurements of inside-negative potentials in fragmented sarcoplasmic reticulum that are generated by an ion exchange process under conditions in which Ca^{++} release, monitored by arsenazo III, is observed. The possible role of the potential gradient in this process is considered.

The behavior of the potential-sensitive probes diS-C_2-(5), diI-C_1-(5), Nile Blue A, oxonol VI, and the oxonol WW781 have been extensively studied by Beeler et al.[117] in sarcoplasmic reticulum vesicles from rabbit skeletal muscle, at least some of which were presumably type II (See the previous pages.). The potential-dependent absorbance signal, monitored by dual-wavelength spectroscopy, was calibrated using diffusion potentials derived from K^+ or Cl^- distributions that were assumed to obey the Nernst relationship. The WW781 dye, however, was insensitive to potential gradients generated in this manner, and, appeared to exhibit optical signals that reflected changes in the surface potential of the sarcoplasmic reticulum vesicles, due to the presence of Ca^{++} or other cations. The diI-C_1-(5) probe exhibited linear responses to diffusion potentials in the range of ± 30 mV from zero potential, whereas, diS-C_2-(5) and oxonol VI signals were linear over − 100 mV and + 100 mV range from zero, respectively. These two probes tended to be significantly less sensitive to inside-positive and inside-negative potentials, respectively. With the exception of WW781, the probes were assumed to distribute across the sarcoplasmic reticulum membrane according to the prevailing membrane potential. No quantitative dye-membrane binding studies were performed. Ivkova et al.,[118] however, have determined constants for the partition of diS-C_3-(5) between various aqueous and membrane phases. In sarcoplasmic reticulum, the critical dye-membrane ratio for probe aggregation was 6 and 9 mol dye per 1000 mol lipid in a sucrose and a salt medium, respectively. Based on these data, a mechanism involving the aggregation of this probe in the sarcoplasmic reticulum membrane has been suggested for the potential-dependent loss of fluorescence yield observed in this preparation.

The absolute magnitude of the probe potential-dependent absorbance signals observed by Beeler et al.[117] was decreased by passively loaded calcium maleate in work with diI-C_1-(5) and diS-C_3-(5) in the presence of inside-negative diffusion potentials, whereas the oxonol VI signal obtained with an inside-positive potential was increased by the presence of Ca^{++} in the range of 0 to 50 mM. The effects due to the presence of Ca^{++} on the probe potential-sensitive signal were essentially the same as those observed during ATP-driven Ca^{++} transport. During active transport, the optical signals of the probes appeared to be roughly proportional to the amount of Ca^{++} accumulated and to persist as long as the Ca^{++} was retained in the vesicle. The latter observation indicates that the probe responses to the direct effect of the ion flux on membrane properties, such as the surface potential cannot be readily distinguished from that due to transmembrane electrical potential gradients. The presence of Ca^{++} was further shown to produce probe optical signals when no membrane potential was present that were qualitatively indistinguishable from the optical responses of the dyes to diffusion potentials, again with the exception of WW781. Nile Blue® A tended to be the least sensitive to the presence of Ca^{++} and appeared to be the most promising of the probes tested as an indicator of negative transmembrane potentials.

By comparing the optical probe response in the sarcoplasmic reticulum system undergoing ATP-driven Ca^{++} transport in the presence of K^+ + valinomycin and in the absence of the ionophore, an estimate of any membrane potential present during active transport could be obtained as the difference in the probe response under these two conditions which were assumed to have the same surface potential component in the probe optical signal. The results with diI-C_1-(5) and oxonol VI suggest that a small inside-negative potential of approximately 10 mV is present during Ca^{++} accumulation, i.e., that Ca^{++} translocation in sarcoplasmic reticulum is electrogenic. Various possible mechanisms for the generation of this negative potential, which could possibly govern Ca^{++} release and accumulation under physiological conditions are discussed.

In related work, Meissner[119] has concluded that Ca^{++} uptake is increased by either a pH gradient or an inside-negative diffusion potential monitored by diO-C_5-(5), but, that the formation of an inside positive potential during ATP-dependent Ca^{++} uptake is prevented by compensating ion translocation in sarcoplasmic reticulum vesicles in order to avoid impeding the uptake of this ion. Based on work with diS-C$_3$-(5) calibrated with diffusion potentials over a ± 100 mV range, Dupont,[120] however, has suggested that in the absence of oxalate, the first few turnovers of the ATPase in sarcoplasmic reticulum vesicles build up an inside-positive potential of approximately 60 mV that is inhibitory to ATPase activity. The apparent electrogenicity of the pump is constant at 7 (± 1) mV, as long as an inside-negative potential is maintained, but, is gradually reduced with increasing vesicle depolarization; the latter observations are in agreement with those of Zimniak and Racker[110] on a reconstituted system. Pechatnikov and Pletnev[121] have also suggested that Ca^{++} transport in sarcoplasmic reticulum is electrogenic, based on the observation that the ATP-driven uptake of this ion increases the rate of dissipation of valinomycin-induced K$^+$ diffusion potentials monitored by diS-C$_3$-(5).

In earlier work with a series of type diO-C$_n$-(3) (n = 3, 4, 5, and 6) and diO-C$_2$-(5) oxycarbocyanine probes, Russell et al.[122] reached similar conclusions to those relevant to the cyanines, i.e., that although the ATP-dependent signals contain a component that is sensitive to the membrane potential, a relatively large portion of these signals is also due to the binding of Ca^{++} and ATP to sarcoplasmic reticulum vesicles. These investigators have also evaluated a number of oxonols derived from barbituric acids, diBA-C$_n$-(2 m + 1), where n = 4, m = 2, and n = 4, m = 1, as well as merocyanines such as M540 and related probes. In contrast to previous work, no evidence could be found in sarcoplasmic reticulum vesicles that the latter dyes were sensitive to changes in diffusion potential across the vesicle membrane. That these probes may respond to changes in electrostatic surface potential caused by the binding of cations was suggested.

Haeyaert et al.[123] have demonstrated that an increase in M540 fluorescence yield in sarcoplasmic reticulum vesicles is correlated with an increment in Ca-ATPase activity and that the dye signal could be reversed by valinomycin or various agents that either inhibit or block Ca^{2+} transport across the vesicle membrane. It was suggested that Ca^{2+} accumulation in this preparation may be electrogenic, but the possibility that the dye is responding to Ca^{2+} diffusion potentials was not excluded. In view of the findings of Russell et al.[84,122] however, regarding the surface potential sensitivity of M540, see previous pages, the behavior of this probe in the sarcoplasmic reticulum vesicle preparation must be interpreted with considerable caution. Peng et al.[124] have compared Ca^{++} uptake and the membrane potential in sarcoplasmic reticulum vesicles from normal and ischemic swine myocardium using murexide and diS-C$_3$-(5), respectively. Both membrane potential and Ca^{++} uptake progressively decreased with the length of the ischemic period, the onset of membrane alterations due to ischemia occurring within 15 min, the shortest ischemic period employed; these membrane alterations could be mimicked in preparations from nonischemic myocardium preincubated with deoxycholate. As also observed by Beeler et al.,[117] a correlation between membrane potential and Ca^{++} uptake apparently exists under these conditions that may be related to sarcoplasmic reticulum dysfunction during ischemic episodes.

Haynes[125] has shown that at low ionic strength, the binding constant describing the association of ANS to phospholipid vesicles is governed by the electrostatic surface potential of the membrane. This property of the probe has been exploited by Chiu et al.[126] in sarcoplasmic reticulum vesicle measurements to obtain the surface potentials on either side of the bilayer at physiological ionic strengths. The binding of the probe to the two membrane surfaces could be distinguished by the time-dependence of this process observed as a fluorescence-yield signal in rapid-mixing work. The association of the probe with the external membrane surface occurred during the mixing time of the stopped-flow apparatus, whereas, the association with the internal membrane surface is governed by the rate at which ANS can permeate the membrane, a process that operates on the time scale of several seconds. Analysis of the fast and slow phase data yielded potentials referenced to the external bathing medium of − 10 to − 15 mV for both the outer and inner membrane surfaces of the Ca^{++}

Table 3

SARCOPLASMIC RETICULUM

Author(s)	Probe	Ref.
Beeler, Farmen, and Martonosi	diS-C$_2$-(5), diI-C$_1$-(5), nile blue, WW781, ox-onol VI	117
Chiu, Mouring, Watson, and Haynes	ANS	126
Dupont	diS-C$_3$-(5)	120
Haeyaert, Verdonck, and Wuytack	merocyanine 540	123
Ivkova, Pechatnikov, Ivkov, and Pletnev	diS-C$_3$-(5)	118
Meissner	diS-C$_3$-(5)	119
Meissner and Young	diO-C$_5$-(3)	112
Ohnishi	diO-C$_2$-(5)	116
Pechatnikov and Pletnev	diS-C$_3$-(5)	121
Pechatnikov, Pletnev, and Rizvanov	diS-C$_3$-(5)	113
Pechatnikov, Rizvanov, Ivkova, Pletnev, and Afanas'ev	diS-C$_3$-(5)	114
Pechatnikov, Rizvanov, and Pletnev	diS-C$_3$-(5)	115
Peng, Straub, and Murphy	diS-C$_2$-(5)	124
Russell, Beeler, and Martonosi	merocyanine 540, diBa-C$_4$-(3), diBa-C$_4$-(5)	84
Russell, Beeler, and Martonosi	diO-C$_3$-(3), diO-C$_4$-(3), diO-C$_5$-(3), diO-C$_6$-(3), diO-C$_2$-(5)	122
Young, Allen, and Meissner	diS-C$_3$-(5)	111
Zimniak and Racker	ANS	110

ATPase-rich sarcoplasmic reticulum fraction and the Ca^{++}-binding protein fraction. A small asymmetry in the surface charge density estimates was detected for the Ca^{++}-binding protein-rich membrane fraction. This asymmetry could be accounted for on the basis of the membrane phospholipid composition; the later conclusion indicates that the acidic Ca^{++}-binding proteins must be located several tens of Å from the membrane surface and hence do not contribute significantly to the surface potential. A mechanism is proposed whereby asymmetric exposure of a membrane surface to Ca^{++} causes a transient difference in surface potential on either side of the bilayer, defined as a macroscopic membrane potential and measurable by electrodes, which is eventually dissipated by ion translocation through the bilayer with the generation of an electrical potential across the membrane, but, with no difference in the two surface potentials remaining. The latter microscopic transmembrane potential is not measurable by electrodes, but, could be sensed by proteins that could undergo conformation changes that regulate the permeability of the sarcoplasmic membrane to Ca^{++}.

VI. MISCELLANEOUS PREPARATIONS

Van Walraven et al.[20] have reconstituted the the ATPase from the thermophylic cyanobacterium *Synechococcus* 6716 into proteoliposomes prepared from native lipids. A detailed study of proton translocation and charge separation in this preparation has been carried out using the native carotenoids and oxonol VI as membrane potential indicators and aminoacridine, neutral red and cresol red as pH indicators. A maximal value of $\Delta\Psi$ of about 130 mV when nigericin is present has been obtained from calibration procedures based on K^+ diffusion potentials. See Reference 87, however, regarding problems in the calibration of oxonol VI with such potentials. On the assumption that neutral red (at low $MgCl_2$ concentration) and aminoacridine are located near the membrane leaflet surfaces, that the carotenoids and oxonol VI are intramembrane probes, and that cresol red and neutral red (at high $MgCl_2$ concentrations) are bulk-phase pH indicators, a model for energy transduction in this preparation has been proposed in which the primary charge separation and proton translocation occurs either within the bilayer and/or near the membrane-bulk-phase interface region. The rationale for this model is a kinetic one based on the observation that the response time to ATP addition of the surface or intramembrane probes is much faster than that of those in the bulk phase, indicating a rather slow equilibration of proton translocation in the bilayer and interfacial regions with the bulk phase, the energy in the latter phase serving as a "buffer" for smoothing out fluctuations in steady-state photosynthetic or respiratory electron flow.

In earlier work, Sone et al.[127] have reconstituted the TF_0-F_1 ATPase from the thermophilic bacterium PS 3 into vesicles consisting of either native phospholipids or soybean lipids. The formation of $\Delta\Psi$, and ΔpH were followed using ANS and 9-aminoacridine fluorescence quenching, respectively. The ANS energy-dependent fluorescence signal was calibrated using K^+ diffusion potentials and the Nernst equation, whereas, the 9-aminoacridine signal was calibrated using the procedure developed by Rottenberg.[128] $\Delta\Psi$ could be enhanced at the expense of ΔpH by TRIS and ΔpH increased at the expense of $\Delta\Psi$ by NO_3^-. A maximal value of $\Delta\mu_H$ of about 250 mV was found under the conditions of this investigation; the latter value is reasonably close to the nominally 300 mV required for a stoichiometry of two protons translocated per ATP hydrolyzed when ATP, ADP, and P_i concentrations are 0.12 mM and the standard free energy of ATP hydrolysis is -8.0 kcal/mol. Using ANS and 9-aminoacridine as probes of $\Delta\Psi$ and ΔpH, respectively, the rate of ATP synthesis as a function of $\Delta\mu_H$ was obtained in a subsequent investigation.[129] A steep dependence of the synthesis was observed when $\Delta\mu_H$ was above 200 mV that appeared to reach a plateau when the ATPase activity was rate-limiting. No such plateau was observed in similar work by Thayer and Hinkle[130] on bovine heart submitochondrial particles when $\Delta\Psi$ was kept constant and ΔpH varied, but, a plateau appeared to develop when $\Delta\Psi$ was varied and ΔpH maintained constant.

Bennett and Spanswick[131] have employed oxonol VI and 9-aminoacridine as probes of the membrane potential and pH gradient, respectively, in ATPase-containing vesicles prepared from the corn root tonoplast membrane. Oxonol VI absorbance was calibrated with K^+ diffusion potentials generated by valinomycin and 9-aminoacridine fluorescence by the method of Deamer et al.[132] In the presence of Cl^-, Mg-ATP addition to this preparation generated an inside-positive membrane potential of 100 mV. The addition of Cl^- salts decreased the potential. In the presence of 50 mM Cl^-, a ΔpH of 1.1 units was established upon addition of Mg-ATP. The kinetics of proton transport were biphasic, consisting of a linear component and a saturable component with a K_m of 4 to 5 mM Cl^-. The decrease in $\Delta\Psi$ was similarliy biphasic, indicating that Cl^- activates electrogenic proton transport as a permeant ion. The saturable component could be abolished by anion channel blockers and was ascribed to Cl^- translocation via a channel, the linear component apparently being due to permeation of the bilayer by Cl^-. That the ATPase may be closely associated with a Cl^- channel and that together they catalyze electroneutral exchange of H^+ and Cl^- was suggested.

Defour et al.[133] have incorporated the ATPase from the yeast *Schizosaccharomyces pombe* into azolectin vesicles by a freeze-thaw procedure. The probe 9-amino-6-chloro-2-methoxyacridine (ACMA) was used to monitor the ΔpH found during ATP hydrolysis. The value of ΔpH was estimated to be 3.6 units from a back titration of the ACMA ATP-dependent fluorescence quenching with HCl. No independent measurements of $\Delta\Psi$ were performed.

Kita et al.[134] have demonstrated that at low oxygen concentrations *Escherichia coli* synthesize cytochrome b_{558}-d complex as an alternate oxidase in order to more efficiently use oxygen at low concentrations, since the K_m for oxygen in the purified b_{558}-d complex is approximately eightfold lower than that for the cytochome b_{562}-o complex. In vesicles containing the reconstituted cytochrome b_{558}-d complex, the substrate ubiquinol-1 induces a quenching of diS-C_3-(5) fluorescence which can be reversed by the uncoupler SF 6847 or the oxidase inhibitors HQNO or zinc sulfate. The cytochrome b_{558}-d complex in reconstituted form can apparently generate an electrochemical gradient across the membrane and appears to function as a coupling site in this bacterial system.

Harikumar and Reeves[135] have investigated the behavior of the ATP-dependent proton pump in a new lysosome preparation from rat kidney cortex. The addition of lysosomes to a diS-C_3-(5) solution causes a quenching of the probe fluorescence which can be partially reversed by agents such as ammonium sulfate, chloroquine, and K^+ + nigericin, all of which collapse the membrane proton gradient; these observations indicate that the lysosomal pH gradient contributes to $\Delta\Psi$. In the presence of valinomycin, the diS-C_3-(5) fluorescence signal increase was a linear function of the log of the K^+ concentration. The addition of Mg-ATP to the preparation caused a rapid increase in the probe-fluorescence yield which could be reversed by protonophores, permeable anions, *N*-ethylmaleimide and K^+ + valinomycin, but, was unaffected by ammonium sulfate, nigericin + K^+, or sodium vanadate. Oligomycin was without effect at concentrations below 2 μg/mℓ but at higher concentrations it partially reversed that ATP-induced signal. At the higher concentrations, however, the inhibitor also affected the probe response to valinomycin-induced diffusion potentials, indicating that it was altering the conductivity of the membrane. The investigators suggest that at concentrations below 2 μg/mℓ, oligomycin interfers with the ability of diS-C_3-(5) to respond to changes in the membrane potential. Based on the distribution of $^{86}Rb^+$ + valinomycin or $S[^{14}C]N^-$, Mg-ATP addition increased the potential of the lysosomal interior by 40 to 50 mV. The lysosomal pump thus appears to operate in an electrogenic fashion.

Cuppoletti and Sachs[136] have used diS-C_3-(5) to compare diffusion potentials in resting and stimulated gastric microsomes from rabbit mucosae. Both preparations developed K^+-gradient-dependent potentials were detected only in the stimulated microsome preparation. It was concluded that activation of acid secretion in stimulated parietal cells may involve at least the appearance of a discrete Cl^- conductance in the proton pump-associated membrane.

Loh et al.[137] have isolated secretory vesicles from the pituitary intermediate lobe, highly purified with respect to lysosomal and mitochondrial content, which process proopiomelanocortin (ACTH/endorphin prohormone). The pH of the interior of these vesicles was estimated to be below 5.6 from measurements using 9-aminoacridine as a probe. The vesicle pH gradient was collapsed by both ammonium sulfate and nigericin +K^+ when the external medium pH was 7. Oxonol VI was used to demonstrate that Mg-ATP addition generated a FCCP-sensitive, inside-positive membrane potential.

Giraudat et al.[138] have solubilized the ATP-dependent proton translocase from chromaffin granule membranes using bile salts and nonionic detergents, this type detergent being essential for an active preparation. When the translocase was reconstituted into vesicles, oxonol V was used to demonstrate that ATP-dependent membrane potentials were formed in this preparation, that was further shown to accumulate norepinephrine in the presence of ATP, suggesting at least partial reconstitution of the catecholamine uptake system.

Ohkuma et al.[139] have employed diS-C_3-(5) in investigations of rat liver lysosomes. These investigators have concluded that at low K^+ concentration levels, a resting membrane potential of 100 to 120 mV, inside-negative, is present that was ascribed primarily to a K^+ gradient, but, with an additional contribution from protons, since the lysosomal membrane is not completely impermeable to the latter species. A membrane potential of approximately 20 to 40 mV was generated by Mg-ATP in this preparation as judged by the partial reversal of the diS-C_3-(5) fluorescence quenching observed upon binding of this probe to the lysosome preparation; the probe response was apparently calibrated using K^+ diffusion potentials with a intralysomal K^+ concentration of 50 mM assumed. The ATP-induced potential, as measured using the probe-fluorescence signal, was roughly coincident with values obtained in the presence of this nucleotide by TPMP and SCN uptake. It was thus concluded that in contrast to the suggestion of Schneider,[140] the lysosomal proton pump is electrogenic.

Bennett et al.[141] have employed diS-C_3-(5) in investigations of the R2 phase of the early receptor electrical potential that apparently develops during the metarhodopsin I to metarhodopsin II transition in rod outer segment membrane vesicles. A flash-induced transient

Table 4
MISCELLANEOUS PREPARATIONS

Author(s)	Probe(s)	Ref.
Bennett, Michel-Villaz, and Dupont	diS-C$_3$-(5)	141
Bennett and Spanswick	oxonol VI	131
Cuppoletti and Sachs	diS-C$_3$-(5)	136
Giraudat, Roisin, and Henry	oxonol V	138
Harikumar and Reeves	diS-C$_3$-(5)	135
Kita, Konishi, and Anraku	diS-C$_3$-(5)	134
Loh, Tam, and Russell	oxonol VI	137
Ohkuma, Moriyama, and Takano	diS-C$_3$-(5)	139
Sone, Yoshida, Hirata, and Kagawa	ANS	129
Sone, Yoshida, Hirata, Okamoto, and Kagawa	ANS	127
Van Walraven, Marvin, Koppenaal, and Kraayenhof	oxonol VI, carotenoids	20

alteration of the dye fluorescence yield was observed to develop on a time scale of tens of milliseconds that was ascribed to the formation of an inside-negative potential; an additional fluorescence yield alteration that develops on the timescale of tens of seconds was ascribed to protein surface charge alterations. The time course of the transient probe signal was similar to that associated with a known protonation change of the protein that occurs during the metarhodopsin transition. It was suggested that the protation change may be involved in the primary charge separation event leading to the formation of the membrane potential. Contributions to the transient dye signal from surface potential alterations were not eliminated, however. The possibility that the transient potential may be involved in the activation of disk membrane enzymes was considered.

VII. SUMMARY REMARKS

The preceding applications of molecular probes have illustrated that in many energy-transducing preparations, estimates of the prevailing potential gradient can be obtained that agree with those obtained by independent methods, such as labeled-permeant ion distribution work.[24,26,27,71] The agreement of membrane potential values obtained with a number of probes, such as the often-employed diS-C$_3$-(5) and safranine probes in mitochondria is particularly noteworthy. The distinctly different viewpoint and membrane potential estimates of Tedeschi and associates,[8,64,65] using not only molecular probes, but also microelectrodes and permeant ion distribution techniques that do not support the presence of a significant membrane potential in mitochondria, however, must be considered in evaluating information obtained with these probes in this system. The current disagreement appears to hinge on the significance of well-accepted osmotic changes observed at the onset of energy transduction upon the highly compartmentalized matrix volume of the mitochondrion and the appropriate volume estimates to be used in determining ion concentration ratios. These problems and associated uncertainties that they engender are common to most if not all estimates of the membrane potential in mitochondria, since the more successful molecular probes appear to distribute across the inner membrane of the mitochondrion in a manner similar to permeant ions. In systems that are unlikely to be subject to large changes in volume at the onset of energy transduction, such as submitochondrial particles or vesicles containing the mito-chondrial ATPase, however, significant membrane potentials have been measured.[6,20,71,95] Tedeschi[8] has noted that the membrane potential does appear to be an intermediate in energy transduction in these cases and has carefully avoided generalizing their findings in intact mitochondria to other preparations.

An alternative to the uncertainty associated with the calibration of energy-dependent probe optical signals with diffusion potentials is the null-point phosphate potential variation method employed by Bashford and Thayer.[71] This method, however, provides the total electrochemical gradient, and to obtain $\Delta\Psi$ an independent method of estimating ΔpH must be available. Akerman and Wikström[26] however, did note that the value of $\Delta\Psi$ as measured by safranine in rat liver mitochondria was a function of the phosphate potential.

The nature of the charge separation event to which molecular probes are sensitive and indeed if they are reliable indicators of transmembrane potential vs. local potentials or changes in surface electrostatic potential is a fundamental issue regarding the application of these indicators and one of the most difficult to answer. The case of ANS, being one of the earliest of molecular probes to be used in membrane systems, has been well-studied. Azzi et al.[67] obtained a value of $\Delta\Psi$ in submitochondrial particles oxidizing succinate that is in agreement with independent measurements, whereas, Ferguson et al.[69] has suggested that this probe senses events at the level of the mitochondrial ATPase, and Haynes[125] and Chiu et al.[126] have recently used this probe in measurements of surface potentials in model and sarcoplasmic reticulum preparations. ANS has also been employed by Robertson and Rottenberg[21] to estimate the surface potential of the mitochondrial inner membrane. These investigators concluded that if a significant change in the surface potential does occur under energy-transducing conditions, as predicted by Malpress,[17] this change is masked by the effect of $\Delta\Psi$ on the ANS optical signal. The uncertainty in what charge gradient(s) this probe senses and under what conditions it is a reliable indicator of $\Delta\Psi$ has stimulated the development of many of the other probes cited in the text. The observation of thiocyanate-sensitive oxonol VI absorption spectrum red shifts in a unreconstituted complex V preparation reported by Kiehl and Hanstein[74] raises the possibility of a local charge separation probe sensitivity in this case, although the interpretation of these results is complicated by uncertainty in whether or not the preparation was effectively vesicular. Based on the work of Bashford and Thayer[71] and Smith and Chance[72] with oxonol VI such problems are at least not yet apparent in the application of this probe to submitochondrial particles. The problem of distinguishing the probe response to a transmembrane potential from that due to changes in surface potential in preparations such as sarcoplasmic reticulum in which massive ion translocation is occurring is particularly acute, as the work of Beeler et al.[117] and Russell et al.[84,122] has demonstrated. The estimates of $\Delta\Psi$ present during Ca^{++} uptake are thus uncertain, with highly variable values being reported by several investigators.[50,119,120] Similar problems in monitoring $\Delta\Psi$ with these indicators during Ca^{++} uptake by mitochondria are probably also present, although the extent of the optical signals from probes such as safranine[45,49,50] or diS-C_3-(5)[27] due to the presence of Ca^{++} under these conditions has not been extensively investigated. On the positive side of this issue, however, is the possibility of employing probes such as merocyanine 540 or perhaps WW781 that do not appear to readily permeate the mitochondrial membrane[80,83,84,117] to investigate surface electrostatic potential changes that may accompany energy transduction, as suggested by the theoretical work of Malpress[17] and Kell.[19]

The location of these probes in the membrane is closely related to the nature of the charge separation processes to which they are sensitive. Since most of these indicators are charged, binding energetics considerations suggest that the portion of the probe molecule bearing the charge will be at or near the membrane-aqueous phase interface; the depth of penetration of the probe chromophore into the membrane, however, is largely unknown. The experiments summarized by Chance[142] in which the location of a number of *N*-phenylnapthylamine type probes has been investigated in model membranes, using proton NMR and paramagnetic perturbants, X-ray diffraction, and resonance energy transfer techniques, is one of the few studies designed to provide such information. Work that is in progress in the author's laboratory indicates that a number of negatively charged oxonols and merocyanine 540 do not affect the chemical shift or relaxation times of the ^{31}P NMR signal from dimysteroyl-phosphatidylcholine vesicles.[143] These observations suggest that these anionic probes may not penetrate the lipid headgroup to the level of the phosphate group, at least in the absence of a potential gradient, possibly due to charge repulsion between the probes and the phosphate moiety which bears a negative charge in this system. Cationic diS-C$_3$-(5) and diS-C$_4$-(5), however, cause a broadening of the ^{31}P lipid resonance and a decrease in both the spin lattice (T$_1$) and spin-spin (T$_2$) relaxation times. These changes appear to be due at least in part to a dye-driven vesicle fusion process, as judged from studies based on electron microscopy.

The question of probe location and orientation in the membrane bilayer is particularly important in the case of electrochromic indicators such as the carotenoids present in photosynthetic systems[99-101] or the charge shift probes synthesized by Loew and associates,[144,145] since these indicators can respond only to the portion of the potential gradient that falls across the probe dimensions. Probes such as merocyanine 540 or WW781 that do not readily permeate the membrane bilayer would be expected to be particularly susceptible to surface potential alterations, a property exploited in the case of ANS by Haynes[125] and Chiu et al.[126] in model and sarcoplasmic reticulum vesicles.

Among the more attractive applications of molecular probes is the possibility of using them to detect initial charge separation events in membranes and to follow the formation of prevailing steady-state potential gradients. The response time of electrodes or classical radioactively labeled ion distribution techniques are likely to be too slow for such studies. The ability of optical indicators to follow such events is closely related to the mechanism responsible for the observed energy-dependent probe signals. Many of the probes that exhibit large, easily detected energy-linked optical signal changes appear to do so by a redistribution process. The time course of the probe signals is often second-order, with the halftime depending on the probe to membrane concentration ratio.[72,80,98] The rate-limiting step under these conditions appears to be the initial transfer of the probe from the bulk aqueous phase or from the unstirred layer[1] to membrane binding sites; a high local dye concentration is required for fast probe response times.[146] The redistribution process can be followed by a slower phase membrane permeation process. At low dye to membrane concentration ratios, where virtually all of the probe is bound, other mechanisms may become important. The disagreement between Peters et al.[95] and Galmiche and Girault[97] over the mechanism of the oxonol VI optical signal in chloroplast preparations appears to be related at least in part to the differing probe to membrane concentration ratios employed in the work of these two groups.

It should also be noted that the membrane potential range over which the response of potential-sensitive redistributing-type probes is linear is also dependent on the binding properties of these indicators.[3,85] Conditions favoring the fastest achievable probe response time and those providing the greatest linearity in the probe response to membrane potential may, unfortunately, be mutually exclusive. The case of safranine[46] is notable in this respect.

The response time of the more commonly employed molecular probes to charge separation in membranes is probably too slow for the detection of the elemental events leading to the formation of steady-state potential gradients. In the chloroplast and chromatophore preparations cited previously in the text,[95,96,98] the response of oxonol VI was found to lag that of the carotenoids to flash excitation. These indicators probably also lag the development of membrane potentials in respiring preparations such as mitochondria[26,28,46] and submitochondrial particles,[69,72] but there are no intrinsic potential-sensitive indicators present in these systems with which to compare the extrinsic probe kinetics. The problem of a lagging probe response time in respiring membrane systems is not as severe as in preparations such as the photosynthetic reaction center, where initial charge separation occurs within 10 psec after illumination,[147] (and references therein) since the turnover numbers of the electron transport chain components suggest a millisecond timescale in this case. Clearly, additional work in the development of fast-responding extrinsic potential-sensitive probes is needed.

The carotenoids present in photosynthetic systems and recently developed charge shift probes[144,145] appear to respond to charge separation by an electrochromic mechanism;[148] see References 22 and 23 however. Since this process operates on the time scale of an electronic transition (10^{-15} sec), it is ideally suited for time-resolved applications of molecular probes. The successful use of β-carotene by Johnson et al.[100] in a reconstituted system is particularly encouraging, since earlier attempts to incorporate carotenoids in mitochondrial preparations have been unsuccessful, possibly because a preferred carotenoid orientation was not assumed in the mitochondrial membrane. A disadvantage of the electrochromic or charge shift probes is the small signals associated with this process. A probe having a (large) permanent dipole moment difference between the ground and an excited state of 20 debye in a field of 10^5 volts/cm, which is typical for biological preparations, undergoes a shift in the absorption or emission spectrum of only 1 to 2 nm. Specialized spectrometers are required to detect such small signals, which can also be easily obscured by other mechanisms, such as redistribution processes if the are operable in the system of interest. Szalontai[101] has circumvented this problem, as well as that of scattering from turbid samples, by the use of Raman spectroscopy, whereas, Ehrenberg et al.[103] have successfully employed the extrinsic probe RH-160 in a microsecond kinetics study of bacteriorhodopsin, again using specialized instrumentation (Figure 3). Loew and associates[144,145] have been developing a series of strongly binding charge shift indicators that exhibit what appear to be electrochromic responses of increasing magnitude to potential gradients in model-membrane preparations. An electrochromic signal from these promising probes in mitochondria or other energy-transducing preparations has, however, thus far not been detected, but much additional work is required with these indicators in energy-transducing systems.

Considerable progress has been made in the application of safranine[48] or other probes, such as oxonol V[6] in measurements in which the probe signals appear to be virtually completely due to changes in the mitochondrial membrane potential in the perfused heart using surface fluorimetry techniques that correct for changes in blood flow and organ motion that could otherwise mask the probe response. A number of probes have been screened for use in monitoring mitochondrial membrane potentials in several whole-cell systems.[56-59] The rhodamine class of dyes appears to be a promising tool for this type of investigation, especially rhodamine 123 (Figure 1). The accumulation of this and similar probes in the matrix of mitochondria can be readily detected using fluorescence microscopy. Thus far, only qualitative measurements of charge gradients across the mitochondrial membrane *in situ* have been attempted. Since the rhodamines have high quantum yields, they can be employed at low concentrations where concentration quenching effects can be minimized as they are accumulated in the matrix of these organelles. Conditions under which the fluorescence yield is linearly proportional to probe concentration may therefore be possible. The probes could then be used as optically detectable permeable ions and the potential calculated from their distribution across the membrane. Loew et al.[149] have presented preliminary work in which rhodamine probes were employed to estimate the membrane potential in intact cells. There is some evidence that several probes, such as oxonol V and diS-C$_3$-(5) are able to sense changes in electrical activity at the level of the mitochondrion in the exposed cerebral cortex of the rat or gerbil; an account of preliminary work with oxonol V has appeared.[6] Work in progress in the author's laboratory indicates that the response of several probes monitored by surface fluorimetry is similar in its temporal development to that of pyridine nucleotide fluorescence yield changes, which originate primarily from the mitochondria, during the onset of anoxia or spreading depression episodes initiated by either topical application of 1 *M* KCl or the injection of KCl or bicuculine, a seizure-inducing drug, into the lateral ventricle. The probe-fluorescence yield goes through a series of cycles in which there is an initial increase in the signal that then reverts to the base line, (Figure 4). This behavior can be eliminated by uncouplers, which dissipate not only the mitochondrial mem-

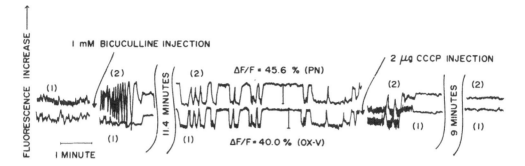

FIGURE 4.   Spreading depression induced by the injection of 10 nmol bicuculline into the lateral ventricle of the mongolian gerbil *(Meriones unguiculatus)*. Traces (1) and (2) are the oxonol and pyridine nucleotide fluorescence signals, respectively. The injection of 2 μg of the uncoupler CCCP into the lateral ventricle completely eliminates the spreading-depression signal. The cortex tissue was stained for approximately 30 sec with a 0.1 mM oxonol V solution in physiological saline. The dye and pyridine nucleotide emission were excited at 580 and 365 nm, respectively; the emission signals were monitored using 2-60 and 3-73 cutoff filters, respectively. The instrument RC time constant was 600 msec. Light-guide diameter: 2 mm. During spreading-depression episodes induced by bicuculline, the change in the 580 nm reflectance signal, which is a measure of blood-volume change contributions to the fluorescence-yield signals, was less than 4%, as measured in separate experiments. (From Evans, D. and Smith, J. C., *Brain Res., 409*, 350, 1987. With permission.)

brane potential but also any plasma membrane potential present. The fluorescence yield cycle can also be eliminated by more specific electron transport inhibitors, antimycin A, rotenone, cyanine, and azide. Subsequent injection of bicuculine will not reinitiate the spreading depression cycle. A preliminary account of these investigations has appeared.[150]

VIII. PRACTICAL CONSIDERATIONS IN THE SELECTION OF OPTICAL PROBES

The tables in this chapter contain a variety of dyes that appear to function as indicators of potential gradients, as well as a small number of intrinsic potential-sensitive probes, mainly carotenoids that are found in photosynthetic systems. Membrane preparations derived from experimental animals, however, often lack intrinsic charge separation indicators; the small ATP-induced cytochrome c oxidase soret band shift signal[10] present in mitochondria being one of the few such probes available in these systems. Intrinsic indicators offer the advantage of being native to the membrane of interest; the introduction of a necessarily perturbing extrinsic probe can be avoided when the former probes are present and can be monitored without interference from the other membrane pigments. A large number of probes of varying optical properties are now available to circumvent potential interference from the membrane system itself when optical measurements are performed; the absorption and emission spectra of the extrinsic probes listed in the tables covers a wavelength range from the near-ultraviolet region to the far-red region of the visible spectrum. The absorption maximum of polyenes increases as the number of carbons in the conjugated carbon chain is incremented,[9] a property that can be useful in designing probes the optical properties of which do not overlap those of intrinsic pigments.

A choice concerning the method(s) of monitoring the probe signal must be made rather early in the probe-selection process. Absorption, fluorescence emission, and to a lesser extent, Raman spectroscopy have all been used either individually or in combination in following the response of probes to charge separation. The nature of the preparation to a considerable degree determines the spectroscopic procedures that can be employed in such experiments. Suspensions of organelles or preparations derived therefrom are usually amenable to either dual-wavelength absorbance of fluorescence emission yield measurements. The former dual-wavelength approach allows for the virtual elimination of Rayleigh scattering from turbid samples but is usually not as sensitive as fluorescence emission measurements. The latter form of spectroscopy is subject to scattering artifacts since the emission is usually monitored at a single wavelength, or scanned spectra are obtained without a reference wavelength for subtractive scattering-correction procedures. The latter problem was overcome by Kauppinen and Hassinen[48] who employed a specialized spectrometer in which the probe fluorescence signal was measured relative to a reflectance signal at the same wavelength as that of the probe emission or at a wavelength isosbestic to but below the fluorescence excitation wavelength. Probe fluorescence can also interfere with Raman spectroscopy measurements.

Single-cell studies[56-59] usually require the fluorescence technique because of the sensitivity advantage that this method offers over absorption spectroscopy. In tissue-level work there is usually no prospect of using transmission/absorption spectroscopy so some emission or reflectance technique must be employed. Surface fluorescence performed using appropriately designed cell holders in commercial instruments or fiber optics methods have been employed in such work.[6,150]

N-arylnapthalenes, such as ANS, MNS, and TNS, exhibit potential sensitivity only in the fluorescence mode, whereas a large number of polyene dyes exhibit both energy-dependent fluorescence and absorption signal changes. The former class of probes are thus not normally employed in work requiring absorption spectroscopy. The relative potential-dependent absorbance and fluorescence signals of the polyene probes are sensitive to changes in substituents that do not affect the optical chromophore. For example, oxonol V, the phenyl derivative, produces small absorption spectrum red shifts (3 to 5 nm) in submitochondrial particles, but, undergoes an uncoupler-sensitive 80% fluorescence-yield quenching in the presence of ATP or substrates. However, the commonly employed propyl derivative, oxonol VI, exhibits a 15 to 20 nm red shift, but, only a 20% fluorescence-yield quenching in this preparation. In systems in which the potential-dependent probe signal originates from a redistribution process, the signal magnitude is usually a function of the probe to membrane concentration ratio, as well as medium ionic strength.[26,70,71,80] The optimum ratio must be determined in each preparation under investigation.

In preparations capable of generating an electrochemical gradient (Equation 1), the relative sensitivity of the probe under consideration to $\Delta\Psi$ vs. ΔpH must be addressed in the selection process if measurements of electrical potential gradients are the goal of the intended investigations. By using electroneutral ionophores such as nigericin or employing NH_4^+, $\Delta p\Psi$ can be increased at the expense of ΔpH. Probes that are specifically sensitive to $\Delta\Psi$ should exhibit an increase in the absolute magnitude of the signal under these conditions.[26,70,71] Even if the probe appears to be specifically potential-sensitive, the possibility of contributions from local charge gradients[69] or from surface potential changes to the probe signal must be considered.[80,117] Surface potential effects can be minimized by performing probe-related observations in media of high ionic strength, e.g., 100 mM monovalent salts, whereas some probes can be used in studies related to surface potentials when the ionic strength is low.[117,125,126]

Quantitative measurements of membrane potential using optical indicators require that they be calibrated, one of the more difficult aspects in the application of molecular probes to energy-transducing preparations. One of the more common procedures is to construct a calibration graph by plotting the probe optical signal (or a parameter proportional to it) as a function of diffusion potentials generated by agents that selectively render the membrane highly permeable to certain ions. The combination of the ionophore valinomycin and K^+ or various protonophores, such as FCCP and H^+ are often used. The Nernst equation can then be used to obtain $\Delta\Psi$:

$$\Delta\Psi = 2.303 \frac{RT}{F} \log \frac{[X]_0}{[X]_i} \tag{3}$$

where [X] is the molar concentration of the permeable species. (Note that Tedeschi[8] has severely questioned the use of this procedure in intact mitochondria.) The latter calculation is based on the assumption that a single permeant species is present. In some cases, the membrane permeability for other ions cannot be neglected, and a more complex relationship such as the Goldman-Katz equation that takes into account the relative permeabilities of the several permeant species must be used; a notable example is the red-cell membrane permeability to Cl^-.

As an example of the procedures used to obtain $\Delta\Psi$ from the Nernst equation, the data of Pick and Avron[85] shown in Figure 2 is used. The data shown in Figure 2(A) was obtained by injecting KCl into a washed chloroplast suspension in the presence of the fluorescent probe OCC and valinomycin and subsequently plotting the percentage fluorescence signal change as a function of the external molar K^+ concentration. At a K^+ concentration of 10^{-2} M, the probe signal is 110% of the control; since the concentration of K^+ in the internal volume of the organelle was estimated to be 5×20^{-4} M, the Nernst equation yields a value of $\Delta\Psi$ of about 48 mV. Note that the factor 2.303 (RT/F) is approximately 60 mV near room temperature. The preceding calculations are repeated point by point to obtain the calibration plot shown in Figure 2(C). The plot shown in Figure 2(B) was obtained by similar calculations in which K^+ was replaced by H^+ in the Nernst equation and FCCP used as the protonophore. Note that in Figure 2(C), the calibration plot exhibits apparent saturation near 175 mV (inside-positive). For probes that respond to $\Delta\Psi$ by redistribution processes, such regions of nonlinear response to $\Delta\Psi$ are common; there is generally a limited range of potentials over which the response of the probe is linear, and this region is dependent on the membrane-binding properties of the probe and the dye to membrane concentration ratio,[46] among other factors. Laris et al.[24] have, for example, found that the response of diS-C$_3$-(5) to K^+ diffusion potentials in intact mitochondria is linear between -25 and -120 mV but tends toward saturation at potentials less than 25 mV (inside-negative). In general, a probe derivative should be selected ideally from a homologous series of dyes, the signal from which is linear over the potential range of interest.

The use of the Nernst equation to obtain probe calibration plots requires that the concentration of the permeant species in the internal volume of the preparation be known. The latter concentration can be predetermined by isolating or incubating the preparation of interest in media of preselected electrolyte concentration. In some cases, however, it is necessary to estimate the actual internal volume of the preparation; such estimates may be made by using radioactively labeled species trapped in the internal volume provided that corrections for binding of these labeled species to the membrane are made. The details of these procedures are discussed by Rottenberg.[128] For mitochondria preparations, the internal volume may be estimated as 1 $\mu\ell$/mg of mitochondrial protein.

The change in medium ionic strength caused by manipulating K^+ or other salt concentrations necessary to generate a range of diffusion potential values can itself alter the interaction of the probe with the membrane.[87,88] Under these conditions, K^+ diffusion potentials cannot be employed in calibration efforts. In some cases, proton gradients generated by acid or base addition to the membrane system under investigation, often in the presence of a protonophore such as FCCP, can be used to obtain probe calibration plots such as that shown in Figure 2(B).

Null-point titration procedures have been used in lieu of[71] or to complement[110] probe calibration efforts based on diffusion potentials. In systems capable of maintaining an electrochemical gradient, null-point titrations involving the phosphate potential provide an equilibrium value of the latter potential that must be converted to $\Delta\mu_H$. This calculation requires a knowledge or at least the assumption of a value for the H^+/ATP stoichiometry. An independent method for separating ΔpH and $\Delta\Psi$ must also be available. In some cases,[71] both potential-sensitive and pH-indicating probes can be employed. If the minimum in both probe responses to alterations in the phosphate potential occurs at the same value of this potential, and, if one of the two components can be independently obtained, the remaining component can be calculated by subtraction. The methods described by Roggenberg[128] can, for example, be used to estimate ΔpH when 9-aminoacridine is employed. Details concerning probe calibration protocols and null-point titrations can be found in the indicated papers[71,87,88] and reviews.[1,3]

The kinetics of the signal development of an indicator in response to charge separation is highly dependent on the mechanism(s) responsible for the probe signal. Since the mechanism(s) may differ greatly among the several membrane preparations, it is difficult to predict how fast a probe will respond to a charge gradient in a given situation. The case of M540 signal-time course in submitochondrial particles[80] and in the squid giant axon[81] is such an example. When rapid response of a probe to charge separation is required, intrinsic electrochromic indicators, such as the carotenoids (Figure 3), or charge-shift extrinsic indicators are the probes of choice provided that the small signals observed, especially with the extrinxic probes, can be detected.[99-101,103] The apparent first-order rate constants characterizing the optical signal development of redistributing-type probes in response to membrane energization increase as the probe concentration is incremented at constant membrane concentration. Rapid response times thus can require substantial concentrations of these probes that may fall outside the linear response range of the calibration curve relating the probe energy-dependent signal and $\Delta\Psi$ and that can cause electron transport inhibition or uncoupling of oxidative phosphorylation. By lowering the membrane concentration to the minimum necessary for an adequate signal to noise ratio, the probe concentration can also be minimized and the fast response time may still be retained.

As previously suggested, virtually any of the commonly employed optical probes, when present at elevated concentrations, can reduce the functional and structural integrity of the membrane preparation under investigation. A common observation is that these probes alter the permeability of the membrane to various ions resulting in an uncoupling effect of the indicator. Certain cyanines, such as diS-C₃-(5) and derivatives thereof[28,32,33] inhibit NADH dehydrogenase-driven electron transport in mitochondrial systems. Probes such as M540 and some cyanines[2,81] also are prone to decomposition under high-intensity illumination conditions, producing degradation products toxic to membranes. In practice, this problem occurs more often in tissue or organ-level studies than when membrane suspensions are involved. Potential-sensitive probes should thus be selected that have an inherently low capacity for uncoupling and inhibitory action in the preparation to be investigated and then employed at the lowest concentrations compatible with linear signal development with $\Delta\Psi$ and/or adequate signal response times.

ACKNOWLEDGMENTS

This work was supported by the Georgia State University Vice President's Research Fund and by U.S. Health and Human Services grants GM30552 and RR 07171. The micrographs shown in Figure 1 were kindly provided by Dr. A. A. Divo.

ABBREVIATIONS USED

ACMA: 9-amino-6-chloro-2-methoxyacridine
ADP: adinosine 5′-diphosphate
ANS: 1-anilino-napthalene-8-sulfonate

AS: 12-(9-anthroyl)stearic acid

ATP: adinosine 5'-triphosphate

CAT_{12}: 4-(dodecyldimethyl ammonium)-1-oxyl-2,2,6,6-tetramethylpiperidine bromide

CC_6: 3,3'-dihexyl-2,2'-oxacarbocyanine

CCCP: carbonyl cyanide *m*-chlorophenylhydrazone

DASPMI: dimethylaminostyrylmethylpyridinium iodide

DCCD: *N,N'*-dicyclohexylcarbodiimide

ΔG_P: phosphate potential

diBA-C_4-(3): bis[1,3-dibutylbarbituric acid-(5)]trimethinoxonol

diBA-C_4-(5): bis[1,3-dibutylbarbituric acid-(5)]pentamethinoxonol

diI-C_1-(5): 3,3'-dimethyl-2,2'-indodicarbocyanine

diO-C_2-(3): 3,3'-diethyl-2,2'-oxacarbocyanine

diO-C_3-(3): 3,3'-dipropyl-2,2'-oxacarbocyanine

diO-C_4-(3): 3,3'-dibutyl-2,2'-oxacarbocyanine

diO-C_5-(3): 3,3'-dipentyl-2,2'-oxacarbocyanine

diO-C_6-(3): 3,3'-dihexyl-2,2'-oxacarbocyanine

diO-C_2-(5): 3,3'-diethyl-2,2'-oxacarbocyanine

diO-C_6-(5): 3,3'-diethyl-2,2'-oxadicarbocyanine

diS-C_2-(5): 3,3'-diethylthiodicarbocyanine

diS-C_3-(5): 3,3'-diethylthiodicarbocyanine

diS-C_4-(5): 3,3'-dibutylthiodicarbocyanine

diS-C_5-(5): 3,3'-dipentylthiodicarbocyanine

DNP: 2,4-dinitrophenol

DSMP: 2-(dimethylaminostyryl)-1-methylpyridinium cation

EDTA: ethylenediaminetetraacetic acid

EGTA: ethyleneglycol bis(β-aminoethylether)-*N,N,N',N'* tetraacetic acid

FCCP: carbonyl cyanide *p*-trifluoromethoxyphenylhydrazone

HQNO: 2-*n*-heptyl-4-hydroxyquinoline-*N*-oxide

ITP: inosine 5'-triphosphate

merocyanine 540(M540, MC-I): 5-[(3-sulfopropyl-2(3H)-benzoxazolylidene)-2-butenyli-
dene]-1,3-dibutyl-2-thiobarbituric acid

MNS: 2-(*N*-methyl-anilino)napthalene-6-sulfonate

NADH: β-nicotinamideadenine dinucleotide

Nbf-Cl (NBD-Cl): 4-chloro-7-nitrobenzofurazan

OCC: 3,3'-dipentyloxacarbocyanine

oxonol V: bis[3-phenyl-5-oxoisoxazol-4-yl]pentamethineoxonol

oxonol VI: bis[3-propyl-5-oxoisoxazol-4-yl]pentamethineoxonol

oxonol VII: bis[3-methyl-5-oxoisoxazol-4-yl]pentamethineoxonol

oxonol VIII: bis[3-heptyl-5-oxoisoxazol-4-yl]pentamethineoxonol

oxonol IX: bis[3-nonyl-5-oxoisoxazol-4-yl]pentamethineoxonol

PHM: pigeon heart mitochondria

P_i: inorganic phosphate, PO_4^{3-}

Q: quinone

RH-160: 4(4'-dibutylaminophenyl-1':3'-diethyl)-1-δ-sulfobutyl-pyridinium hydroxide

SF6847: 3,5-di-*tert*-butyl-4-hydroxybenzylidenemalononitrile

TNS: toluidinonapthalenesulfonate

$TPMP^+$: methyl triphenylphosphonium ion

TPP: tetraphenylphosphonium ion

TRIS: tris(hydroxymethyl)aminomethane

WW781: 1,3-dibutylbarbituric acid (5)-1-(*P*-sulfophenyl)-3-methyl,5 pyrazolone
pentamethinoxonol

REFERENCES

1. **Waggoner, A. S.**, Optical probes of membrane potential, *J. Membr. Biol.*, 27, 317, 1976; **Waggoner, A. S.**, Dye indicators of membrane potential, *Annu. Rev. Biophys. Bioeng.*, 8, 47, 1979.
2. **Cohen, L. B. and Salzberg, B. M.**, Optical measurements of membrane potential, *Rev. Physiol. Biochem. Pharmacol.*, 83, 35, 1978.
3. **Bashford, C. L. and Smith, J. C.**, The use of optical probes to monitor membrane potential, in *Methods in Enzymology*, Vol. 55, Fleischer, S. and Packer, L., Eds., Academic Press, New York, 1979, 569.
4. **Bashford, C. L.**, The measurement of membrane potential using optical indicators, *Biosci. Rep.*, 1, 183, 1981.
5. **Chance, B., Baltscheffsky, M., Vanderkooi, J., and Cheng, W.**, Localized and delocalized potentials in biological membranes, in *Perspectives in Membrane Biology*, Estrado-O, S. and Gitler, S., Eds., Academic Press, New York, 1974, 329.
6. **Bashford, C., Barlow, C., Chance, B., Smith, J., Silberstein, B., and Rehncrona, S.**, Some properties of the extrinsic probe oxonol v in tissues, in *Frontiers of Biological Energetics*, Vol. 2, Scarpa, A., Dutton, P. L., and Leigh, J. S., Eds., Academic Press, New York, 1979, 1305.
7. **Smith, J., Powers, L., Prince, R., Chance, B., and Bashford, L.**, Potential sensitive oxonol dyes: model systems to organelles, in *Frontiers of Biological Energetics*, Vol. 2, Scarpa, A., Dutton, P. L., and Leigh, J. S., Eds., Academic Press, New York, 1979, 1293.
8. **Tedeschi, H.**, The mitochondrial membrane potential, *Biol. Rev.*, 55, 171, 1980.
9. **Platt, J. R.**, Wavelength formulas and configuration interaction in Brooker dyes and chain molecules, *J. Chem. Phys.*, 25, 80, 1956.
10. **Wikström, M. K. F. and Saari, H. T.**, Conformational changes in cytochrome aa_3 and ATP-synthetase of mitochondrial membranes and their role in mitochondrial energy transduction, *Mol. Cell. Biochem.*, 11, 17, 1976.
11. **Mitchell, P.**, Coupling of phosphorylation to electron and hydrogen transfer by a chemi-osmotic type of mechanism, *Nature (London)*, 191, 144, 1961.
12. **Mitchell, P. and Moyle, J.**, Stoichiometry of proton translocation through the respiratory chain and adeonisine triphosphate systems of rat liver mitochondria, *Nature (London)*, 208, 147, 1965.
13. **Mitchell, P. and Moyle, J.**, Estimation of membrane potential and pH difference across the cristae membrane of rat liver mitochondria, *Eur. J. Biochem.*, 7, 471, 1969.
14. **Mitchell, P.**, Possible molecular mechanisms of the proton motive function of cytochrome systems, *J. Theor. Biol.*, 62, 327, 1976.
15. **Boyer, P. D.**, Correlation of the binding change mechanism with a new concept for proton translocation and energy transduction, in *H^+-ATPase (ATP Synthase): Structure, Function, Biogenesis. The F_0F_1 Complex of Coupling Membranes*, Papa, S., Ed., ICSU Press and Adriatica Editrice, Bari, 1984, 329.
16. **Slater, E. C., Berden, J. P., and Herweijer, M. A.**, A hypothesis for the mechanism of respiratory-chain phosphorylation not involving the electrochemical gradient of protons as obligatory intermediate, *Biochim. Biophys. Acta*, 811, 217, 1985.
17. **Malpress, F. H.**, A coulombic hypothesis of mitochondrial oxidative phosphorylation, *J. Theor. Biol.*, 109, 501, 1984.
18. **Woelders, H., van der Zande, W. J., Colen, A.-M. A. F., Wanders, R. J. A., and van Dam, K.**, The phosphate potential maintained by mitochondria in state 4 is proportional to the proton-motive force, *FEBS Lett.*, 179, 278, 1985.
19. **Kell, D. B.**, On the functional proton current pathway of electron transport phosphorylation: an electrodic view, *Biochim. Biophys. Acta*, 549, 55, 1979.
20. **Van Walraven, H. S., Marvin, H. J. P., Koppenaal, E., and Kraayenhoff, R.**, Proton movements and electric potential generation in reconstituted ATPase proteoliposomes from the thermophilic cyanobacterium *Synechococcus* 6716, *Eur. J. Biochem.*, 144, 555, 1984.
21. **Robertson, D. E. and Rottenberg, H.**, Membrane potential and surface potential in mitochondria: fluorescence and binding of 1-anilinonapthalene-8-sulfonate, *J. Biol. Chem.*, 258, 11039, 1983.
22. **Matsuura, K., Masamoto, K., Itoh, S., and Nishimura, M.**, Effect of surface potential on the intramembrane electrical field measured with carotenoid spectral shift in chromatophores from *Rhodopseudomonas sphaeroides*, *Biochim. Biophys. Acta*, 547, 91, 1979.
23. **Masamoto, K., Matsuura, K., Itoh, S., and Nishimura, M.**, Membrane-potential- and surface-potential-induced absorbance changes of merocyanine dyes added to chromatophores from *Rhodopseudomonas sphaeroides*, *Biochim. Biophys. Acta*, 638, 108, 1981.
24. **Laris, P. C., Bahr, D. P., and Chaffee, R. R. J.**, Membrane potentials in mitochondrial preparations as measured by means of a cyanine dye, *Biochim. Biophys. Acta*, 376, 415, 1975.
25. **Harris, E. J. and Pressman, B. C.**, The direction of polarity in the mitochondrial transmembrane potential, *Biochim. Biophys. Acta*, 172, 66, 1969.

26. **Akerman, K. O. and Wikström, M. K. F.**, Safranine as a probe of the mitochondrial membrane potential, *FEBS Lett.*, 68, 191, 1976.
27. **Zinchenko, V. P., Holmuchamedov, E. L., and Evtodienko, Yu. V.**, Determination of membrane potential in different metabolic states of mitochondria by means of a fluorescent dye, *Stud. Biophys.*, 72, 91, 1978.
28. **Bammel, B. P., Brand, J. A., Germon, W., and Smith, J. C.**, The interaction of the extrinsic potential-sensitive molecular probe diS-C₃-(5) with pigeon heart mitochondria under equilibrium and time-resolved conditions, *Arch. Biochem. Biophys.*, 244, 67, 1986.
29. **Bashford, C. L. and Smith, J. C.**, The determination of oxonol binding parameters by spectoscopic methods, *Biophys. J.*, 25, 81, 1979.
30. **Packer, L.**, Metabolic and structural states of mitochondria. I. Regulation by adenosine diphosphate, *J. Biol. Chem.*, 235, 242, 1960.
31. **Packer, L. and Tappel, A. L.**, Light scattering changes linked to oxidative phosphorylation in mitochondrial membrane fragments, *J. Biol. Chem.*, 235, 525, 1960.
32. **Connover, T. E. and Schneider, R. F.**, Interaction of certain cationic dyes with the respiratory chain of rat liver mitochondria, *J. Biol. Chem.*, 256, 402, 1981.
33. **Pechatnikov, V. A., Rizvanov, F. F., and Turchina, S. L.**, Studies of mitochondrial transmembrane potential using cyanine dyes, *Biofizika*, 24, 178, 1979.
34. **Montecucco, C., Pozzan, T., and Rink, T.**, Dicarbocyanine fluorescent probes of membrane potential block lymphocyte capping, deplete cellular ATP and inhibit respiration of isolated mitochondria, *Biochim. Biophys. Acta*, 552, 552, 1979.
35. **Aiuchi, T., Matsunaga, M., Nakaya, K., and Nakamura, Y.**, Effects of probes of membrane potential on metabolism in synaptosomes, *Biochim. Biophys. Acta*, 843, 20, 1985.
36. **Quintanilha, A. T. and Packer, L. B.**, Surface potential changes on energization of the mitochondrial inner membrane, *FEBS Lett.*, 78, 161, 1977.
37. **Quintanilha, A. T., Packer, L., Davies, J. M. S., Racanelli, T. L., and Davies, K. J. A.**, Membrane effects of vitamin E deficiency: bioenergetic and surface charge density studies of skeletal muscle and liver mitochondria, *Ann. N.Y. Acad. Sci.*, 393, 32, 1982.
38. **Bakeeva, L. E., Kirillova, G. P., Kolesnikova, O. V., Konoshenko, G. I., and Mokhova, E. N.**, Influence of palmitate on the energy coupling in lymphocyte mitochondria, *Biochemistry (USSR)*, 50, 650, 1985.
39. **Michejda, J., Adamski, J., Hejnowicz, M., and Hryniewiecka, L.**, Changes in membrane potential in amoeba mitochondria monitored by the fluorescent dye diS-C₃-(5), in *Function and Molecular Aspects of Biomembrane Transport, Developments in Bioenergetics and Biomembranes*, Vol. 3, Quagliariello, E., Palmieri, F., Papa, S., and Klingenberg, M., Eds., Elsevier/North Holland, New York, 331, 1979.
40. **Adamski, J., Morkowski, J., and Michejda, J. W.**, Energization of mitochondria of *Acanthamoeba Castellanii* measured by the fluorescent cyanine dye 3,3′-dipropylthiodicarbocyanide iodide, *Sci. Lett. Poznan*, 20, 39, 1980.
41. **Moore, A. L.**, An evaluation of H⁺ transport *via* the alternative pathway in mung bean mitochondria, in *Functions of Alternative Terminal Oxidases*, Vol. 49, Degn, H., Lloyd, D., and Hill, S. C., Eds., Pergamon Press, Oxford, 141, 1978.
42. **Laris, P. C.**, Evidence for the electrogenic nature of the ATP-ADP exchange in rat liver mitochondria, *Biochim. Biophys. Acta*, 459, 110, 1977.
43. **Lauquin, G. J. M., Villiers, C., Michejda, J., Brandolin, G., Boulay, F., Cesarini, R., and Vignais, P. V.**, Current approaches to the mechanism of ADP/ATP transport across the mitochondrial membrane, in *The Proton and Calcium Pumps, Developments in Bioenergetics and Biomembranes*, Vol. 2, Azonne, G. F., Avron, M., Metcalfe, J. C., Quagliariello, E., and Siliprandi, N., Eds., Elsevier/North Holland, New York, 1978, 251.
44. **Nicholls, D. G.**, The influence of respiration and ATP hydrolysis on the proton electrochemical gradient across the inner membrane of rat liver mitochondria as determined by ion distribution, *Eur. J. Biochem.*, 50, 305, 1974.
45. **Akerman, K. E. O.**, Changes in membrane potential during calcium ion influx and efflux across the mitochondrial membrane, *Biochim. Biophys. Acta*, 502, 359, 1978.
46. **Zanotti, A. and Azonne, G. F.**, Safranine as membrane potential probe in rat liver mitochondria, *Arch. Biochem. Biophys.*, 201, 255, 1980.
47. **Wilson, S. B.**, Energy conservation by the plant mitochondrial cyanide-insensitive oxidase Some additional evidence, *Biochem. J.*, 190, 349, 1980.
48. **Kauppinen, R. A. and Hassinen, I. E.**, Monitoring of mitochondrial membrane potential in isolated perfused rat heart, *Am. J. Physiol.*, 247, H508, 1984.
49. **Riley, W. W., Jr. and Pfeiffer, D. R.**, Rapid and extensive release of Ca²⁺ from energized mitochondria induced by EGTA, *J. Biol. Chem.*, 261, 28, 1986.

50. **Cockrell, R. S.**, Ruthenium red — insensitive Ca^{2+} uptake and release by mitochondria, *Arch. Biochem. Biophys.*, 243, 70, 1985.

51. **Kim, Y. V., Kudzina, L. Yu., Zinchenko, V. P., and Evtodienko, Y. V.**, Chlortetracyclin-mediated continuous Ca^{2+} oscillation in mitochondria of digitonin-treated *Tetrahymena pyriformis*, *Eur. J. Biochem.*, 153, 503, 1985.

52. **Bereiter-Hahn, J.**, Dimethylaminostyrylmethylpyridiniumiodide (DASPMI) as a fluorescent probe for mitochondria in situ, *Biochim. Biophys. Acta*, 423, 1, 1976.

53. **Mewes, H.-W. and Rafael, J.**, The 2(dimethylaminostyryl)-1-methylpyridinium cation as indicator of the mitochondrial membrane potential, *FEBS Lett.*, 131, 7, 1981.

54. **Rafael, J. and Nicholls, D. G.**, Mitochondrial membrane potential monitored in situ within isolated guinea pig brown adipocytes by a styryl pyridinium fluorescent indicator, *FEBS Lett.*, 170, 181, 1984.

55. **Korchak, H. M., Rich, A. M., Wilkenfeld, C., Rutherford, C. E., and Weissman, G.**, A carbocyanine dye, di0-C$_6$-(5), acts as a mitochondrial probe in human neutrophils, *Biochem. Biophys. Res. Commun.*, 108, 1495, 1982.

56. **Johnson, L. V., Walsh, M. L., and Chen, L. B.**, Localization of mitochondria in living cells with rhodamine 123, *Proc. Natl. Acad. Sci. U.S.A.*, 77, 990, 1980.

57. **Johnson, L. V., Walsh, M. L., Bockus, B. J., and Chen, L. B.**, Monitoring of relative mitochondrial membrane potential in living cells by fluorescence microscopy, *J. Cell Biol.*, 88, 526, 1981.

58. **Davis, S., Weiss, M. J., Wong, J. R., Lampidis, T. J., and Chen, L. B.**, Mitochondrial and plasma membrane potentials cause unusual accumulation and retention of rhodamine 123 by human breast adenocarcinoma-derived MCF-7 cells, *J. Biol. Chem.*, 260, 13844, 1985.

59. **Divo, A. A., Geary, T. G., Jensen, J. B., and Ginsburg, H.**, The mitochondrion of *Plasmodium falciparum* visualized by rhodamine 123 fluorescence, *J. Protozool.*, 32, 442, 1985.

60. **Kovac, L. and Varecka, L.**, Membrane potentials in respiring and respiration-deficient yeasts monitored by a fluorescent dye, *Biochim. Biophys. Acta*, 637, 209, 1981.

61. **Mikeš, V. and Dadák, V.**, Berberine deriatives as cationic fluorescent probes for the investigation of the energized state of mitochondria, *Biochim. Biophys. Acta*, 723, 231, 1983.

62. **Mikeš, V. and Yaguzhinskij, L. S.**, Interaction of fluorescence berberine alkyl derivatives with respiratory chain of rat liver mitochondria, *J. Bioenerg.*, 17, 23, 1985.

63. **Tedeschi, H.**, Absence of a metabolically induced electrical potential across the mitochondrial semipermeable membrane, *FEBS Lett.*, 59, 1, 1975.

64. **Tedeschi, H.**, Mitochondrial membrane potential: evidence from studies with a fluorescent probe, *Proc. Natl. Acad. Sci. U.S.A.*, 71, 583, 1974.

65. **Kinnally, K. W., Tedeschi, H., and Maloff, B. L.**, The use of dyes to estimate the electrical potential of the mitochondrial membrane, *Biochemistry*, 17, 3419, 1978.

66. **Williams, W. P., Layton, D. B., and Johnson, C.**, An analysis of the binding of fluorescence probes in mitochondrial systems, *J. Membr. Biol.*, 33, 21, 1977.

67. **Azzi, A., Gherardini, P., and Santato, M.**, Fluorochrome interaction with the mitochondrial membrane. The effect of energy conservation., *J. Biol. Chem.*, 246, 2035, 1971.

68. **Barrett-Bee, K. and Radda, G. K.**, On the nature of the energy-linked quantum yield change in anilinonapthalene sulphonate fluorescence in submitochondrial particles, *Biochim. Biophys. Acta*, 267, 211, 1972.

69. **Ferguson, S. J., Lloyd, W. J., and Radda, G. K.**, On the nature of the energised state of submitochondrial particles: investigations with N-aryl napthalene sulphonate probes, *Biochim. Biophys. Acta*, 423, 174, 1976.

70. **Smith, J. C., Russ, P., Cooperman, B. S., and Chance, B.**, Synthesis, structure determination, spectral properties, and energy-linked spectral responses of the extrinsic probe oxonol v in membranes, *Biochemistry*, 15, 5094, 1976.

71. **Bashford, C. L. and Thayer, W. S.**, Thermodynamics of the electrochemical proton gradient in bovine heart submitochondrial particles, *J. Biol. Chem.*, 252, 8459, 1977.

72. **Smith, J. C. and Chance, B.**, Kinetics of the potential-sensitive extrinsic probe oxonol vi in beef heart submitochondrial particles, *J. Membr. Biol.*, 46, 255, 1979.

73. **Bashford, C. L., Chance, B., Smith, J. C., and Yoshida, T.**, The behavior of oxonol dyes in phospholipid dispersions, *Biophys. J.*, 25, 63, 1979.

74. **Kiehl, R. and Hanstein, W. G.**, ATP-dependent spectral response of oxonol vi in an ATP-P$_i$ exchange complex, *Biochim. Biophys. Acta*, 766, 375, 1984.

75. **Smith, J. C., Hallidy, L., and Topp, M. R.**, The behavior of the fluorescence lifetime and polarization of oxonol potential sensitive extrinsic probes in solution and in beef heart submitochondrial particles, *J. Membr. Biol.*, 60, 173, 1981.

76. **Chance, B., Barlow, C. H., Nakase, Y., Takeda, H., Mayevsky, A., Fischetti, R., Graham, N., and Sorge, J.**, Heterogeneity of oxygen delivery in normoxic and hypoxic states: a fluorimeter study, *Am. J. Physiol.*, 235, H809, 1978.

77. **Sanadi, D. R., Pringle, M., Kantham, L., Hughes, J. B., and Srivastava, A.**, Evidence for the involvement of coupling factor B in the H$^+$ channel of the mitochondrial ATPase, *Proc. Natl. Acad. Sci. U.S.A.*, 81, 1371, 1984.

78. **Pringle, M. J. and Sanadi, D. R.**, Effects of cadmium on ATP driven membrane potential in beef heart mitochondrial proton translocating ATPase: a study using the voltage sensitive dye oxonol vi, *Membr. Biochem.*, 5, 225, 1984.

79. **Wagner, R., Andreo, C., and Junge, W.**, Evidence for a sequestered solvent space in chloroplast ATPase, *Biochim. Biophys. Acta*, 723, 123, 1983.

80. **Smith, J. C., Graves, J. M., and Williamson, M.**, The interaction of the potential-sensitive molecular probe merocyanine 540 with phosphorylating beef heart submitochondrial particles under equilibrium and time-resolved conditions, *Arch. Biochem. Biophys.*, 231, 430, 1984.

81. **Cohen, L. B., Salzberg, B. M., Davila, H. V., Ross, W. N., Landowne, D., Waggoner, A. S., and Wang, C. H.**, Changes in axon fluorescence during activity: molecular probes of membrane potential, *J. Membr. Biol.*, 19, 1, 1974.

82. **Salama, G. and Morad, M.**, Optical probes of membrane potential in heart muscle, *J. Physiol. (London)*, 292, 267, 1979.

83. **Krasne, S.**, Interactions of voltage-sensing dyes with membranes. III. Electrical properties induced by merocyanine 540, *Biophys. J.*, 44, 305, 1983.

84. **Russell, J. T., Beeler, T., and Martonosi, A.**, Optical probe responses on sarcoplasmic reticulum Merocyanine and oxonol dyes, *J. Biol. Chem.*, 254, 2047, 1979.

85. **Pick, U. and Avron, M.**, Measurement of transmembrane potentials in *Rhodopseudomonas Rubrum* chromatophores with an oxycarbocyanine dye, *Biochim. Biophys. Acta*, 440, 189, 1976.

86. **Schuurmans, J. J., Casey, R. P., and Kraayenhof, R.**, Transmembrane electrical potential formation in spinach chloroplasts, *FEBS Lett.*, 94, 405, 1978.

87. **Admon, A., Shahak, Y., and Avron, M.**, Adenosine triphosphate-generated electric potentials in chloroplasts, *Biochim. Biophys. Acta*, 681, 405, 1982.

88. **Shahak, Y., Admon, A., Avron, M.**, Transmembrane electrical potential formation by chloroplast ATPase complex (CF$_1$-CF$_0$) proteoliposomes, *FEBS Lett.*, 150, 27, 1982.

89. **Shahak, Y. and Pick, U.**, A time lag in the onset of ATP-P$_i$ exchange catalyzed by purified ATP-synthase (CF$_0$-CF$_1$) proteoliposomes and by chloroplasts, *Arch. Biochem. Biophys.*, 223, 393, 1983.

90. **DeWolf, F. A., Groen, B. H., van Houte, L. P. A., Peters, F. A. L. J., Krab, K., and Kraayenhof, R.**, Studies on well-coupled photosystem I-enriched subchloroplast vesicles. Neutral red as a probe for external surface charge rather than internal protonation, *Biochim. Biophys. Acta*, 809, 204, 1985.

91. **Junge, W., Ausländer, W., McGeer, A. J., and Runge, T.**, The buffering capacity of the internal phase of thylakoids and the magnitude of the pH change inside under flashing light, *Biochim. Biophys. Acta*, 546, 121, 1979.

92. **Schuurmans, J. J., Peters, F. A. L. J., Leeuwerik, F. J., and Kraayenhof, R.**, On the association of electrical events with the synthesis and hydrolysis of ATP in photosynthetic membranes, in *Vectorial Reactions in Electron and Ion Transport in Mitochondria and Bacteria*, Palmieri, F., Quagliariello, E., Siliprandi, N., and Slater, E. C., Eds., Elsevier/North Holland, New York, 359.

93. **Thieg, S. M. and Junge, W.**, The effect of low concentrations of uncouplers on the detectability of proton deposition in thylakoids. Evidence for subcompartmentation and preexisting pH differences in the dark, *Biochim. Biophys. Acta*, 723, 294, 1983.

94. **Junge, W., Hong, Y. Q., Qian, L. P., and Viale, A.**, Cooperative transient trapping of photosystem II protons by the integral membrane portion (CF$_0$) of chloroplast ATP-synthase after mild extraction of the four-subunit catalytic part (CF$_1$), *Proc. Natl. Acad. Sci. U.S.A.*, 81, 3078, 1984.

95. **Peters, F. A. L. J., van der Pal, R. H. M., Peters, R. L. A., Vredenberg, W. J., and Kraayenhof, R.**, Studies on well-coupled photosystem-I-enriched subchloroplast vesicles; Discrimination of flash-induced fast and slow electric potential components, *Biochim. Biophys. Acta*, 766, 169, 1984.

96. **Peters, F. A. L. J., van Spanning, R., and Kraayenhof, R.**, Studies on well coupled photosystem-I-enriched subchloroplast vesicles; Optimization of ferredoxin-mediated cyclic photophosphorylation and electric potential generation, *Biochim. Biophys. Acta*, 724, 159, 1983.

97. **Galmiche, J. M. and Girault, G.**, ATP hydrolysis and membrane potential in spinach chloroplasts, *FEBS Lett.*, 118, 72, 1980.

98. **Bashford, C. L., Chance, B., and Prince, R. C.**, Oxonol dyes as monitors of membrane potential. Their behavior in photosynthetic bacteria, *Biochim. Biophys. Acta*, 545, 46, 1979.

99. **Koyama, Y., Long, R. A., Martin, W. G., and Carey, P. R.**, The resonance raman spectrum of carotenoids as an intrinsic probe for membrane potential. Oscillatory changes in the spectrum of neurosporene in the chromatophores of *Rhodopseudomonas Sphaeroides*, *Biochim. Biophys. Acta*, 548, 153, 1979.

100. **Johnson, J. H., Lewis, A., and Gogel, G.**, Kinetic resonance raman spectroscopy of cartenoids: a sensitive kinetic monitor of bacteriorhodopsin mediated membrane potential changes, *Biochem. Biophys. Res. Commun.*, 103, 182, 1981.

101. **Szalontai, B.**, Light induced membrane potential changes in *Halobacterium halobium* observed with high time resolution by resonance Raman spectroscopy, *Biochem. Biophys. Res. Commun.*, 100, 1126, 1981.

102. **Stockenius, W., Lozier, R. H., and Bogomolni, R. A.**, Bacteriorhodopsin and the purple membrane of halobacteria, *Biochim. Biophys. Acta*, 505, 215, 1978.

103. **Ehrenberg, B., Meiri, Z., and Loew, L. M.**, A microsecond kinetic study of the photogenerated membrane potential of bacteriorhodopsin with a fast responding dye, *Photochem. Photobiol.*, 39, 199, 1984.

104. **Packam, N. K., Greenwood, J. A., and Jackson, J. B.**, Generation of membrane potential during photosynthetic electron flow in chromatophores from *Rhodopseudomonas capsulata*, *Biochim. Biophys. Acta*, 592, 130, 1980.

105. **Webster, G. D., Cogdell, R. J., and Lindsay, J. G.**, Identification of the carotenoid present in the B-800-850 antenna complex from *Rhodopseudomonas capsulata* as that which responds electrochromically to transmembrane electric fields, *Biochim. Biophys. Acta*, 591, 321, 1980.

106. **Matsuura, K., Masamoto, K., Itoh, S., and Nishimura, M.**, Surface potential of the periplasmic side of the photosynthetic membrane of *Rhodopseudomonas sphaeroides*, *Biochim. Biophys. Acta*, 592, 121, 1980.

107. **Symons, M., Nuyten, A., and Sybesma, C.**, On the calibration of the carotenoid band shift with diffusion potentials, *FEBS Lett.*, 107, 10, 1979.

108. **Cirillo, V. P. and Gromet-Elhanan, Z.**, Steady-state measurements of ΔpH and $\Delta \Psi$ in *Rhodopseudomonas rubrum* chromatophores by two different methods, *Biochim. Biophys. Acta*, 636, 224, 1981.

109. **Kell, D., Ferguson, S., and John, P.**, Measurements by a flow dialysis technique of the steady state proton motive force in chromatophores from *Rhodopseudomonas rubrum*, *Biochim. Biophys. Acta*, 502, 111, 1978.

110. **Zimniak, P. and Racker, E.**, Electrogenicity of Ca^{2+} transport catalyzed by the Ca^{2+}-ATPase from sarcoplasmic reticulum, *J. Biol. Chem.*, 253, 4631, 1978.

111. **Young, R. C., Allen, R., and Meissner, G.**, Permeability of reconstituted sarcoplasmic reticulum vesicles. Reconstitution of the K^+, Na^+ channel, *Biochim. Biophys. Acta*, 640, 409, 1981.

112. **Meissner, G. and Young, R. C.**, Proton permeability of sarcoplasmic reticulum vesicles, *J. Biol. Chem.*, 255, 6814, 1980.

113. **Pechatnikov, V. A., Pletnev, V. V., and Rizvanov, F. F.**, Permeability of the membrane of the sarcoplasmic reticulum for monovalent cations, *Biofizika*, 28, 710, 1983.

114. **Pechatnikov, V. A., Rizvanov, F. F., Ivkova, M. N., Pletnev, V. V., and Afanas'ev, V. N.**, Study of sarcoplasmic reticulum membrane permeability by potential sensitive fluorescent probes, *Biofizika*, 25, 1075, 1980.

115. **Pechatnikov, V. A., Rizvanov, F. F., and Pletnev, V. V.**, Effect of potential sensitive fluorescent dyes on sarcoplasmic reticulum membrane permeability, *Stud. Biophys.*, 93, 95, 1983.

116. **Ohnishi, S. T.**, A method for studying the depolarization-induced calcium ion release from fragmented sarcoplasmic reticulum, *Biochim. Biophys. Acta*, 587, 121, 1979.

117. **Beeler, T. J., Farmen, R. H., and Martonosi, A. N.**, The mechanism of voltage-sensitive dye responses on sarcoplasmic reticulum, *J. Membr. Biol.*, 62, 113, 1981.

118. **Ivkova, M. N., Pechatnikov, V. A., Ivkov, V. G., and Pletnev, V. V.**, Mechanism of the fluorescent response of carbocyanine probe diS-C_3-(5) to membrane potential change, *Biofizika*, 28, 160, 1983.

119. **Meissner, G.**, Calcium transport and monovalent cation and proton fluxes in sarcoplasmic reticulum vesicles, *J. Biol. Chem.*, 256, 636, 1981.

120. **Dupont, Y.**, Electrogenic calcium transport in the sarcoplasmic reticulum membrane, in *Cation Flux Across Biomembranes*, Mukohata, Y. and Packer, L. Eds., Academic Press, New York, 141.

121. **Pechatnikov, V. A. and Pletnev, V. V.**, The effect of active and passive Ca^{2+} transport in the sarcoplasmic reticulum, *Biofizika*, 29, 438, 1984.

122. **Russell, J. T., Beeler, T., and Martonosi, A.**, Optical probe responses on sarcoplasmic reticulum. Oxycarbocyanines, *J. Biol. Chem.*, 254, 2040, 1979.

123. **Haeyaert, P., Verdonck, F., and Wuytack, F.**, Fluorescence changes of the potential-sensitive merocyanine 540 during Ca transport in sarcoplasmic reticulum, *Arch. Int. Pharmacodyn. Ther.*, 244, 333, 1980.

124. **Peng, C. F., Straub, K. D., and Murphy, M. L.**, Alterations of membrane potential and Ca^{2+} flux of sarcoplasmic reticulum vesicles in ischemic myocardium, *Ann. Clin. Lab. Sci.*, 13, 511, 1983.

125. **Haynes, D. H.**, 1-Anilino-8-napthalenesulfonate: a fluorescent indicator of ion binding and electrostatic potential on the membrane surface, *J. Membr. Biol.*, 17, 341, 1974.

126. **Chiu, V. C. K., Mouring, D., Watson, B. D., and Haynes, D. H.**, Measurement of surface potential and surface charge densities of sarcoplasmic reticulum membranes, *J. Membr. Biol.*, 56, 121, 1980.

127. **Sone, N., Yoshida, M., Hirata, H., Okamoto, H., and Kagawa, Y.**, Electrochemical potential of protons in vesicles reconstituted from purified proton-translocating adenosine triphosphatase, *J. Membr. Biol.*, 30, 121, 1976.

128. **Rottenberg, H.**, The measurement of transmembrane electrochemical proton gradients, *Bioenergetics*, 7, 61, 1975.

129. **Sone, N., Yoshida, M., Hirata, H., and Kagawa, Y.**, Adeonosine triphosphate synthesis by electrochemical proton gradient in vesicles reconstituted from purified adeonosine triphosphatase and phospholipids of thermophilic bacterium, *J. Biol. Chem.*, 252, 2956, 1977.

130. **Thayer, W. S. and Hinkle, P. C.**, Kinetics of adenosine triphosphate synthesis in bovine heart submitochondrial particles, *J. Biol. Chem.*, 250, 5336, 1975.

131. **Bennett, A. B. and Spanswick, R. M.**, Optical measurements of ΔpH and $\Delta\Psi$ in corn root membrane vesicles: kinetic analysis of Cl$^-$ effects on a proton-translocating ATPase, *J. Membr. Biol.*, 71, 95, 1983.

132. **Deamer, D. W., Prince, R. C., and Crofts, A. R.**, The response of fluorescent amines to pH gradients across liposome membranes, *Biochim. Biophys. Acta*, 274, 323, 1972.

133. **Dufour, J.-P., Goffeau, A., and Tsong, T. Y.**, Active proton uptake in lipid vesicles reconstituted with the purified yeast plasma membrane ATPase, *J. Biol. Chem.*, 257, 9365, 1982.

134. **Kita, K., Konishi, K., and Anraku, Y.**, Terminal oxidases of *Escherichia coli* aerobic respiratory chain II. Purification and properties of cytochrome b_{558}-d complex from cells grown with limited oxygen and evidence of branched electron carrying systems, *J. Biol. Chem.*, 259, 3375, 1984.

135. **Harikumar, P. and Reeves, J. P.**, The lysosomal proton pump is electrogenic, *J. Biol. Chem.*, 258, 10403, 1983.

136. **Cuppoletti, J. and Sachs, G.**, Regulation of gastric acid secretion via modulation of a chloride conductance, *J. Biol. Chem.*, 259, 14952, 1984.

137. **Loh, Y. P., Tam, W. W. H., and Russell, J. T.**, Measurement of ΔpH and membrane potential in secretory vesicles isolated from bovine pituitary intermediate lobe, *J. Biol. Chem.*, 259, 8238, 1984.

138. **Giraudat, J., Roisin, M.-P., and Henry, J.-P.**, Solubilization and reconstitution of the ATP dependent proton translocase of bovine chromaffin granule membrane, *Biochemistry*, 19, 4499, 1980.

139. **Ohkuma, S., Moriyama, Y., and Takano, T.**, Electrogenic nature of lysosomal proton pump as revealed with a cyanine dye, *J. Biochem. (Tokyo)*, 94, 1935, 1983.

140. **Schneider, D. L.**, ATP-Dependent acidification of intact and disrupted lysosomes. Evidence for an ATP-driven proton pump, *J. Biol. Chem.*, 256, 3858, 1981.

141. **Bennett, N., Michel-Villaz, M., and Dupont, Y.**, Cyanine dye measurement of a light-induced transient membrane potential associated with the metarhodopsin II intermediate in rod-outer-segment membranes, *Eur. J. Biochem.*, 111, 105, 1980.

142. **Chance, B.**, Electron transport and energy-dependent responses of deep and shallow probes of biological membranes, in *Energy Transducing Mechanisms*, Vol. 3, Racker, E., Ed., University Park Press, Baltimore, 1975, 1.

143. **Smith, J. C. and Brand, J.**, The effect of potential-sensitive molecular probes on the ^{31}P NMR spectrum in dimysteroylphosphatidylcholine vesicles, *Biophys. J.*, 45, 331a, 1984.

144. **Loew, L. M. and Simpson, L. L.**, Charge shift probes of membrane potential. A probable electrochromic mechanism for p-aminostrylpyridinium probes on a hemispherical lipid bilayer, *Biophys. J.*, 34, 353, 1981.

145. **Fluhler, E. and Loew, L. M.**, Membrane binding, flip-flop, and voltage dependence spectroscopy of charge shift probes, *Biophys. J.*, 45, 167a, 1984.

146. **Smith, J. C., Frank, S. J., Bashford, C. L., Chance, B., and Rudkin, B.**, Kinetics of the association of potential sensitive dyes with model and energy transducing membranes: implications for fast probe response times, *J. Membr. Biol.*, 54, 127, 1980.

147. **Dutton, P. L., Prince, R. C., Tiede, D. M., Petty, K. M., Kaufmann, K. J., Netzel, T. L., and Rentzepis, P. M.**, Electron transfer in the photosynthetic reaction center, in *Chlorophyll-Proteins, Reaction Centers, and Photosynthetic Membranes*, Vol. 28, Olson, J. M. and Hind, G., Eds., Brookhaven National Laboratory, Upton, N.Y., 1977, 213.

148. **Bucher, H., Wiegand, J., Snavley, B. B., Beck, K. H., and Kuhn, H.**, Electric field induced changes in the optical absorption of a merocyanine dye, *Chem. Phys. Lett.*, 3, 508, 1969.

149. **Loew, L. M., Ehrenberg, B., Fluhler, E., Wei, M., and Burnham, V.**, A search for Nernstian dyes to measure membrane potential in individual cells, *Biophys. J.*, 49, 308a, 1986.

150. **Smith, J. C. and Evans, D.**, The behavior of potential-sensitive molecular probes in the exposed cerebral cortex of the gerbil, *Biophys. J.*, 49, 95a, 1986.

INDEX

A

Printed and bound by CPI Group (UK) Ltd, Croydon, CR0 4YY

22/10/2024

01777630-0005